Construction Supply Chain Management:
Concepts and Case Studies

Dedication

This book is dedicated to Karen

Construction Supply Chain Management:
Concepts and Case Studies

Edited by

Stephen Pryke

Senior Lecturer in Construction and Project Management
Bartlett School of Graduate Studies
University College London

A John Wiley & Sons, Ltd., Publication

This edition first published 2009
© 2009 Blackwell Publishing Ltd

Blackwell Publishing was acquired by John Wiley & Sons in February 2007. Blackwell's publishing programme has been merged with Wiley's global Scientific, Technical, and Medical business to form Wiley-Blackwell.

Registered office
John Wiley & Sons Ltd, The Atrium, Southern Gate, Chichester, West Sussex, PO19 8SQ, United Kingdom

Editorial offices
9600 Garsington Road, Oxford, OX4 2DQ, United Kingdom
350 Main Street, Malden, MA 02148-5020, USA

For details of our global editorial offices, for customer services and for information about how to apply for permission to reuse the copyright material in this book please see our website at www.wiley.com/wiley-blackwell.

The right of the author to be identified as the author of this work has been asserted in accordance with the Copyright, Designs and Patents Act 1988.

Library of Congress Cataloging-in-Publication Data
Construction supply chain management : concepts and case studies / edited by Stephen Pryke.
        p.   cm.—(Innovation in the built environment)
  Includes bibliographical references and index.
    ISBN 978-1-4051-5844-2 (hardback : alk. paper)   1. Construction industry—Management—Case studies.   2. Business logistics—Management—Case studies.   I. Pryke, Stephen.

  TH438.C6425   2009
  690.068′7—dc22

                                                                                    2008047436

A catalogue record for this book is available from the British Library.

Set in 10/12 pt Sabon by SNP Best-set Typesetter Ltd., Hong Kong
Printed and bound in Malaysia by Vivar Printing Sdn Bhd

1   2009

Innovation in the Built Environment

Series advisors

Carolyn Hayles, *Queen's University, Belfast*
Richard Kirkham, *University of Manchester*
Andrew Knight, *Nottingham Trent University*
Stephen Pryke, *University College London*
Steve Rowlinson, *The University of Hong Kong*
Derek Thompson, *Heriot Watt University*
Sara Wilkinson, *University of Melbourne*

Innovation in the Built Environment (IBE) is a new book series for the construction industry published jointly by the Royal Institute of Chartered Surveyors and Wiley-Blackwell. It addresses issues of current research and practitioner relevance and takes an international perspective, drawing from research applications and case studies worldwide.

- presents the latest thinking on the processes that influence the design, construction and management of the built environment

- based on strong theoretical concepts and draws on both established techniques for analysing the processes that shape the built environment – and on those from other disciplines

- embraces a comparative approach, allowing best practice to be put forward

- demonstrates the contribution that effective management of built environment processes can make

Published and forthcoming books in the IBE series

Akintoye & Beck, *Policy, Finance & Management for Public-Private Partnerships*
Lu & Sexton, *Innovation in Small Professional Practices in the Built Environment*
Boussabaine, *Risk Pricing Strategies for Public-Private Partnerships*
Kirkham & Boussabaine, *Whole Life-Cycle Costing*
Booth, Hammond, Lamond & Proverbs, *Solutions to Climate Change Challenges in the Built Environment*

We welcome proposals for new, high quality, research-based books which are academically rigorous and informed by the latest thinking; please contact Stephen Brown or Madeleine Metcalfe.

Stephen Brown
Head of Research
RICS
12 Great George Street
London SW1P 3AD
sbrown@rics.org

Madeleine Metcalfe
Senior Commissioning Editor
Wiley-Blackwell
9600 Garsington Road
Oxford OX4 2DQ
mmetcalfe@wiley.com

Contents

Contributors

Dr Stephen Pryke is a Senior Lecturer in Construction and Project Management, Bartlett School of Graduate Studies, UCL, and is Director of the Masters course Project and Enterprise Management.

Mike Bresnen is Professor of Organisational Behaviour at the University of Leicester School of Management. He is a founding member and Associate Fellow of the IKON (Innovation, Knowledge and Organisational Networking) research centre at Warwick Business School and an Associate Editor of Organisation. He has researched and published widely on the organisation and management of the construction process, as well as on project-based learning, innovation and knowledge management in a variety of other project-based, inter-organisational settings.

Dr Andrew Edkins is a Senior Lecturer in the Bartlett School at UCL and is the course director for the Masters in Strategic Management of Projects. He has researched, taught and been involved in the management of complex projects, concentrating latterly in those using forms of project finance or involving public and private partnerships.

Richard Fellows is a Professor at The University of Hong Kong. He is joint coordinator of CIB W112 'Culture in Construction'. Richard is a member of the editorial boards of several leading journals and has published extensively in journals, conferences and books. He has taught and researched at several UK universities and is visiting professor in China.

Mohieddin Grada was awarded his PhD from Nottingham Trent University and has taught and carried out post-doctoral studies in the field of financial evaluation of construction organisations. He has published a number of papers in his field.

Dr Andrew King is Supply Chain Development Manager for Morgan Ashurst, one of the largest construction contractors in the UK. Andrew's academic background, specialising in construction procurement, is evidenced in numerous publications.

Dr Andrew Knight is a Principal Lecturer and Research Coordinator in the School of Architecture, Design and Built Environment at Nottingham Trent University. Andrew's research interests include construction economics, procurement, property management, sociology of the professions and ethics and professionalism. He is a reviewer for *Construction Management* *and Economics* and on the editorial panel for the Research Paper Series for the Royal Institution of Chartered Surveyors.

Roy Morledge is Professor of Construction Procurement and Associate Dean at the School of Architecture, Design and Built Environment at Nottingham Trent University. He has authored and edited a number of books and academic journal papers in the areas of procurement, value and financial management. Professor Morledge is a referee for *Construction Management and Economics* and on the Editorial Board of the *Journal of Construction Procurement.*

Martin Pitt is Commercial Director for Morgan Ashurst plc.

Keith Potts is a Senior Lecturer in the School of Engineering and Built Environment at the University of Wolverhampton and Award Leader for the RICS accredited MSc Construction Project Management programme. He is author of *Construction Cost Management: learning from case studies* published by Taylor and Francis in 2008. He was previously Senior Quantity Surveyor for Kier Ltd and the Hong Kong Mass Transit Railway Corporation.

Dr Bernard Rimmer is a consultant to The Concrete Centre, and former Visiting Professor at the Department of Construction Management of The University of Reading. During his many years as General Manager of Development and Construction at Slough Estates, he was Chairman of the Design Build Foundation, a member of the Egan Task Force, a Director of Reading Construction Forum, and a Director of the Foundation for the Built Environment (BRE parent).

Martin Skitmore is Professor in the Faculty of the Built Environment and Engineering, Queensland University of Technology, Brisbane. He is a visiting professor at the Hong Kong City University and a member of the editorial boards of a number of leading journals. His interests include cost modeling, affordable housing and commercial management of construction projects.

Dr Hedley Smyth is a Senior Lecturer in the Bartlett School, UCL and Director of Studies for the Postgraduate Research programme. He is author of several books. He has worked in construction and developed marketing capabilities for a number of design practices.

Preface

I am often asked by both students and practitioners what is meant by supply chain management. The term, like so many before it, has entered the everyday language of both researcher and practitioner. I hope that this book will provide an agenda for discussion for the experienced researcher and practitioner and an introduction for the novice. The seeds of this book were sown in my early research career during my time working with British Airports Authority and Slough Estates as well as a number of major developers and construction firms. Those early attempts to emulate manufacturing left a great impression upon me. Yet over the intervening period relatively little has been written about the subject of supply chain management (SCM) in construction.

I am convinced of the importance of SCM whether as a source of tools and techniques in projects or as a higher level theoretical framework for conceptualising the activities of our project organisations.

I hope that this book will be read by undergraduate and postgraduate students of construction, project management, engineering and architecture, as well as quantity surveyors. I hope to show through these pages, and through the discussion of both concepts and practice, that SCM has an important part to play in both academic discussion and practice.

I feel privileged to have worked with a prestigious team of academics and practitioners on this project. I hope you enjoy the final product as much as I enjoyed the journey that led up to the production of this book.

Stephen Pryke
London, November 2008

Acknowledgements

I would like to thank all of the authors that contributed to this book on supply chain management. Technology has enabled us all to collaborate intensively and effectively and yet several of us have never had the opportunity to meet – I hope that this will be remedied soon. Special thanks go to Dr Bernard Rimmer, ex-Construction Manager from Slough Estates plc, not only for contributing to the book but for starting me off in pursuit of the study of supply chains. During his time with Slough Estates he very kindly and generously encouraged an enthusiastic, not so young, PhD student in his studies on project management and social networks. Special thanks also to Dr Hedley Smyth, a contributor to this book and a very valued colleague at UCL. Hedley gave very generously of his time on this and several previous collaborative projects and was very supportive in this supply chain management project. Thanks also to Karen Rubin for her help with editing and the production of graphics; and to Ella Sivyer for her help with editing and the organisation of the production of the individual chapters.

Finally my thanks go to the Bartlett School of Graduate Studies, University College London, which as my employer allowed me the time to work on this project and provided a stimulating environment in which to work and flourish as an academic and a human being.

Introduction

Stephen Pryke

1.1   Supply Chain Management – What Is It?

Any discussion of construction supply chain management (SCM) is usually informed by a wide range of definitions. This diversity of definitions and understanding presents a challenge, which this book addresses. We explore a wide range of conceptual issues that help us to understand the nature of supply chains in construction. In addition, there is case-study material reflecting the work of some of the leading proponents of SCM in UK construction. The premise for this book is the move from the *project* and its management, as the main focus for the management of the construction process, towards the *supply chain* and its management as the main focus. The supply chain is the focus for more effective ways of creating value for clients; as a vehicle for innovation and continuous improvement, integration of systems and perhaps even improved, industry-wide, profitability levels.

Value creation is increasingly viewed as a process facilitated through a supply chain – a network of relationships within which firms are positioned. New and Westbrook (2004) suggest that firms in supply chains must build networks so as to provide complementariness between inner and external abilities, that is to say, effective supply chains need to be supported in networks that extend beyond the immediate linkages of exchange in order to create the value in each link. In the same way that individuals are drawn to, or naturally seek, other individuals with skills, knowledge and attributes that they themselves lack, firms are drawn to form collaborative relationships with other firms with skills, knowledge, attributes and perhaps resources that are complementary to the first firm. Just as individuals in society find it difficult to survive isolated from others, isolates in business are vulnerable and may fail in time unless they possess a unique skill or talent which gives them market power (for example a monopoly supplier or oligopoly of few suppliers in a market of buoyant demand).

The term *supply chain* implies a linear process. This linearity, however, exists only at a high level of abstraction. At an applied level, when we

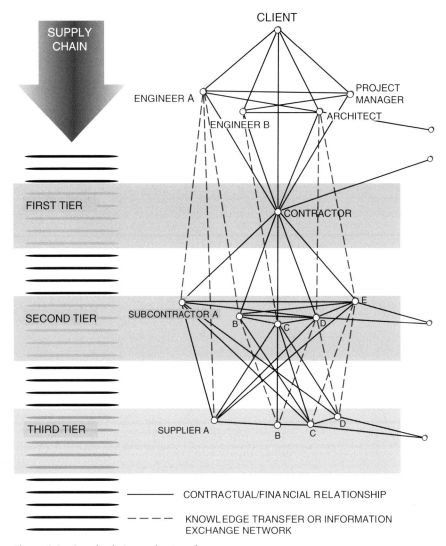

Figure 1.1   Supply chains and networks.

explore the nature and operation of supply management, there is limited linearity, clusters of suppliers coming together in series of dyadic exchanges. Social and market exchanges create social and technical systems which, once in place, are observed as dynamic networks of relationships. The juxtaposition of these two aspects of supply chains is shown in Figure 1.1.

In construction we observe clients, consultants, contractors and suppliers in the broadest sense positioned as *nodes* connected by linkages comprising knowledge transfer, information exchange, directions and financial and contractual relationships. These networks are transitory (Pryke and Smyth, 2006) and the flows are iterative (Pryke, 2001); like neural networks the *nodes* are continually linking and disconnecting depending on the project function to be performed. Each linkage involves flows that produce a response and generate a succession of dyadic or multi-directional flows until a particular function is satisfied and specific issues are resolved.

During the early stages of post-Latham (1994) reform in UK construction, major clients were working hard to find better ways to procure and manage construction services. Partnering had real application for the large, experienced clients that had the resources to experiment with innovative systems. British Airports Authority and Slough Estates plc provided corporate and developer client examples of what the industry was striving for during the mid-1990s. The large number of arguments in support of less adversarial relations and partnering arrangements in supply chains have been advocated post-Latham and post-Egan (DETR, 1998), particularly for the large client. And yet the down side of partnering can potentially be a recipe for complacency on the part of service providers (which might include consultants, contractors, subcontractors and material and component suppliers); and higher outturn costs for clients. Large clients were not slow to realise the vulnerable position that they individually and their organisations found themselves in as a result of abandoning the 'comfort zone' of traditional competitive price bidding on a contract-by-contract basis.

BAA and Slough Estates, among others, started to think about the solution to maintaining value for money and ensuring continuous improvement in the services that both organisations procured in great quantity each year. Bedtime reading for ambitious young executives at BAA during the mid-1990s was *The Machine that Changed the World* (Womack *et al.*, 1990). This seminal work did not deal with construction at all – it referred to a post-mass-production motor manufacturing industry and embraced and expounded lean thinking. The existence of long-term supplier relationships and the relatively intense management of these relationships were central. Major construction clients began to realise that partnering provided a threat and an opportunity – the threat of escalating costs and poor performance from service providers, but the opportunity to collaborate and integrate within the context of long-term relationships. The construction design and production process has some activities that non-construction manufacturers would recognise immediately – standard and semi-standard components incorporated into a system to provide for example heating, ventilation, lift installations or perhaps suspended ceilings; these components are delivered to site and fixed together to form a system or sub-system within the building. The management of the design and supply of standard and semi-standard components, assembled at the final assembly place (in our case, the site) enables the principles found in manufacturing to be fairly simply applied to these parts of the construction scheme. Yet the construction team has to deal with an additional, slightly less ordered type of process, which might be referred to as *site crafted work*. Figure 1.2 illustrates this point through an adaptation of Harland's (1996) work.

A slightly uneasy relationship has developed between construction project management and SCM. Indeed, senior staff at BAA during the mid-1990s insisted that project management as a discipline was 'dead' and that all future reference would be to SCM. The problem remained, and still persists, of how the site-crafted element of new buildings is managed – SCM in its

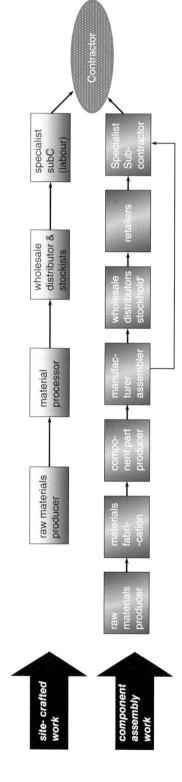

Figure 1.2   An inter-business construction project supply chain. source: adapted from Harland (1996: s67), cited in Pryke (2001).

purest form does not really help us to procure and organise brickwork and plastering, for example, the so-called 'wet trades'. Observers (Green, 2006, for example) might dismiss SCM as the current fad or fashion. Yet, casual enquiry with many construction firms confirms a commitment to SCM. But so often organisations have simply changed the titles given to their procurement staff using the words 'supply chain managers'. SCM involves intensive management activity from a central position within the construction coalition or network either by the client organisation, or by another organisation acting as the client's agent. The time consuming and demanding activities comprising effective supply chain management in construction involve everything 'from the quarry to the finished project'. It requires capacity and knowledge at the centre and a certain commitment from client organisations to make it effective. Recent research (unpublished as this book went to press) carried out at UCL and involving a number of major UK construction client organisations, indicated that the examination of sub-supply chain elements to investigate the incremental accumulation of both cost and value proved extremely instructive. Identifying and mapping small sections of the overall supply chain for a construction project and isolating the costs and value added by each actor's involvement provides the basis for a more enlightened understanding of cost and value in construction projects.

The aim of this book is to demonstrate that SCM in construction is more than a management fad and provides the opportunity for substantial improvements in client and stakeholder value and/or reductions in overall costs. The book brings together the conceptual ideas of some of the best thinkers in academia, with the first-hand experiences of a number of practitioners in construction. It is offered in response to a number of enquiries from academics, students and practitioners about the nature of SCM in construction.

The book is intended to stimulate debate around the subject of SCM in construction, whether the reader regards SCM as a new way of conceptualising the management of projects or whether SCM is regarded as a new technique to be learned. Either way, the reader might benefit from acquiring an understanding of some of the basics of SCM before turning to the discussion within the material included here. Ayers (2004), Copacino (1997), Dyer (1996) and Ptak and Schragenheim (2000) all provide some introductory basics in a non-construction context. For sources dealing with the implementation of SCM in construction see CBPP (2003), Cain (2003), Cox and Ireland (2002), Holti, *et al.* (2000), London and Kenley (2001) and Virhoef and Koskela (2000). At the same time the Project Management Institute Body of Knowledge (PMI, 2000) provides a context from which to understand the relevance of the debate that follows in the various papers offered here.

1.2   Supply Chain Management and Project Management

Project management has evolved through several stages of development, each adding complementary understanding to the existing bodies of knowledge (Pryke and Smyth, 2006):

- *Traditional* project management approach – techniques and tools for application (for example Turner, 1999; cf. Turner and Mûller, 2003; and Koskela, 1992; 2000), which tend to have a production or assembly orientation focused upon efficiency.
- *Functional* management approach – strategic, 'front-end' management of projects (Morris, 1994; cf. Morris and Pinto, 2004), for example programme and project strategies, and partnering (Egan, 1998) and supply chain management (see Green and May, 2003) and other task-driven agendas which dovetail with the traditional approach, for example the waste elimination application of lean production (Koskela, 1992; 2000).
- *Information processing* approach – technocratic input–output model of managing projects (Winch, 2002).
- *Relationship* approach – project performance and client satisfaction, achieved through an understanding of the way in which a range of relationships between people, between people and firms, and between firms as project actors operate and can be managed.

Relations are context specific (Pryke and Smyth, 2006). There are different contexts for relationships, which operate at different levels:

- Business-to-business or organisation-to-organisation;
- Organisation-to-individual representing the business: market and other societal relations (see Gummesson, 2001);
- Individual-to-individual: personal and social relations.

Personal relations can be characterised as (Pryke and Smyth, 2006):

- Authority: management and leadership;
- Task related: function and role;
- Acquaintance: social obligation;
- Friendship: social bonding and reciprocation;
- Sense of identity: who you are (not what you do), such as inheritance and societal recognition – through ownership in business for example.

Organisational relations can be characterised as:

- Individual or personal: the individual represents the organisation;
- Systematic or procedural: personal relations have been enshrined into an approach or systematic way of proceeding in order that the essence of a relationship is replicated at a general level in the future through social or legal obligation (cf. Wenger, 1998);
- Strategies and culture help guide the context in which systems operate, guiding the thinking and behaviour of individuals in order that relations through individuals and systems are aligned;
- Structure of an organisation both reflects relations and governs relations through hierarchy, function and proximity.

The quality of relationships is a key element in the success of a project. The quality may be the product of a range of factors and therefore a consequence of a whole series of dynamic issues. In this way a project team is merely the recipient of those relationships and how they develop both within the project team and with those who are externally feeding into the project. However, relationships are also *managed*. This book is devoted to the management of relationships through the framework of supply chains. The study of projects and their supply chains provides an appropriate context for the analysis of construction projects and programmes within a partnering framework.

1.3   Origins of SCM in Construction

The aim of this section is to review the historical context and nature of SCM in the UK construction industry. SCM is inextricably linked with partnering but whether partnering creates the need for SCM or vice versa is a debatable point. It is clear, however, that the discussion of SCM needs to be preceded by a definition of partnering. As the aim of this book is to deal primarily with SCM, the discussion of partnering is deliberately brief.

1.3.1   A brief history of partnering in construction

> *'Partnering involves two or more organisations working together to improve performance through agreeing mutual objectives, devising a way of resolving any disputes and committing themselves to continuous improvements, measuring progress and sharing gains' Egan (DETR, 1998, page 9)*

Bovis (now Bovis Lend Lease) is credited as being the first UK construction organisation to be involved in a partnering arrangement (Loraine, 1994). The other partner was Marks and Spencer, the retailer, and the arrangement was not called partnering at the time. Given the repetitive nature *of internal fit-outs*, with standardised fittings and finishes, this would have been a situation where Banwell's (Ministry of Public Building and Works 1964) *serial contracting* could have been usefully applied. Loraine (1994), on the other hand, suggests that modern partnering has its origins in the Japanese motor manufacturing of the 1960s and 1970s. The US construction industry began to use partnering in the 1980s, commencing with Shell Oil and Parsons SIP in 1984. The characteristics of these partnering arrangements appear to have been long term relationships between manufacturers and key suppliers, and often included maintenance as well as initial installation (similar in concept to the operation of the lift installation sector of the UK construction industry).

The process of selecting a contractor on the basis of lowest competitive tender is at the heart of what Winch (2000) describes as the *professional system*. Winch identifies the way that activities facing the highest uncertainty in the design stages are insulated from the market by the employment of a consultant, reimbursed on a (non-performance related) fee basis. The industry has seen the evolution of *control actors* (quantity surveyors and clerks of works principally) whose role it was to regulate the activities that remained subject to market forces (Winch, 2000).

Despite some moves towards reform, (notably Design and Build, and management contracting) the industry developed during the latter part of the twentieth century into a low-trust system, in which consultants spent too much of their time ensuring that their professional indemnity insurance cover was not exposed to unnecessary risk, and contractors and their sub-contractors adopted opportunistic behaviour as a means of recovering from unacceptably low tendered profit margins in a context of inappropriately allocated project risks. The industry did not necessarily perceive the context and systems prevailing within UK construction during this period as a problem (Pryke, 2001). Both the Latham and Egan Reports referred to the importance of partnering in reforming the construction industry. The CRINE project in the North Sea (http://www.crine-network.com/, cited in Winch 2000) demonstrated the benefits of partnering in the offshore gas and oil industries, which were related to the mainstream construction industry. The motivation for the introduction of partnering, in this case, was related to the need to drive down costs in order to exploit resources that would otherwise have been unprofitable. The partnering initiative was seen as a vehicle for intensive financial management of the supply chain.

Other, relatively early, examples of UK construction partnering include an example given by Daniels (1991, cited in Betts and Wood Harper, 1994) where a UK brick supplier re-engineered its links with architectural buyers through innovative use of IT. The broad principles of trust and maximisation of each participant's resources and expertise have become the main focus of partnering agreements used within the industry. The Latham Report expressed some concern about the possibility of 'cosy relationships' and offered trust and openness, along with 'mutually agreed and measurable targets for productivity' as possible antidotes to the opportunism with which the industry had become all too familiar (Latham 1994).

Barlow *et al.* (1997) suggest that there are three main perspectives on partnering:

- A construction process, performance-enhancing tool which draws upon synergy and the maximisation of the effectiveness of each participant's resources (Barlow *et al.*, 1997).
- A management process involving strategic planning to improve efficiency in large projects, or perhaps a variant of total quality management. Others have emphasised the *common goals* aspect and partnering as an aid to *collaboration*.
- The *non-contractual governance of construction projects school:* Barlow *et al.*, (1997) referred to 'putting the handshake back into doing business' implying a move towards trust and informal arrangements (or what is referred to as 'keeping the contract in the drawer').

The unique division of labour operating in the construction industry involving designers, contractors and materials suppliers has been cited as a central theme and a focus for reform (see Higgin and Jessop, 1965;

Cherns and Bryant, 1984; and Bresnen, 1997, for example). The construction project coalition is a temporary coalition of firms (see Winch, 1989 and 2000). Each firm represents a discrete (contractually defined) role, and when these roles work together, we hope that partnering will modify the roles and the relationships between them. We might, therefore, regard the construction project as a *role system* (Simon, 1976). A number of the perceived benefits from partnering arise from the ability of this system of roles to improve organisational learning.

Partnering is seen by some as a means of removing these artificial divisions, yet the evidence of the effects on actor roles and relationships is difficult to locate (Bresnen, 2000). There are also varying views about the precise role that contracts and charters play in partnering. One group (notably Quick, 1994; ACTIVE, 1996; Green and McDermott, 1996; cited in Bresnen, 2000) asserts that partnering agreements prevail over the building contract conditions, because of the improved understanding arising out of cross-disciplinary communications. Others (notably Loraine, 1994, referred to above and Roe, 1996, cited in Bresnen, 2000) regard contractual forms of governance as an essential safety net in the event that partnering might fail.

The move towards informality in the governance of construction projects – the move away from *contract management* towards *relationship management*, has bought with it a demand for methods of effectively managing these new types of linkages between project actors. One of these initiatives was supply chain management.

1.3.2   Supply chain management

The aim of this section is to introduce the concept of supply chain management (SCM), establish some definitions and look at the history. SCM in construction is arguably a more recent innovation than partnering. Whilst the Latham Report was unequivocal in its recommendation to the industry to adopt 'partnering' (Latham, 1994, discussed above), the thoughts of the authors were slightly less well developed in relation to SCM. The report contains a reference to a presentation by Dr Bernard Rimmer of Slough Estates plc at a conference organised by 'Contract Journal' and CASEC in December 1993, where the position of the client in relation to the motor-manufacturing industry and the construction industry are compared (Latham, 1994).

The Egan report (DETR, 1998) was quite specific in its reference to SCM. The report recommends the adoption of the following features of SCM:

- Acquisition of new suppliers through *value* based sourcing;
- Organisation and management of the supply chain to maximise innovation, learning and efficiency;
- Supplier development and measurement of suppliers' performance;
- Management of workload to match capacity and incentivisation of suppliers to improve performance;
- Capture of suppliers' innovations in components and systems.

1.3.3   Some definitions of supply chain management

Partnering in its most simple form asks little more of the project actors other than co-operation. Arguably, the industry should be encouraged to abandon the futile pursuit of adversarial and non-collaborative relationships within the context of a system that will never deliver the *customer delight* to which the construction industry's clients aspire and are so often frustrated in achieving.

Howell *et al.* (1996) argue that partnering should be used to facilitate major process re-engineering rather than easing the difficulties encountered in inappropriate systems for the procurement of construction work. This realignment should focus upon the needs of a *concurrent design and production process.*

In order to achieve an output, this re-engineered process must include the management of the various actors in the product supply chain. Views differ as to the nature of this supply chain and it is arguable whether a complex network of organisations working together in a number of non-trade related clusters, are best described as *chains* and this was referred to at the beginning of this chapter. Figure 1.2 illustrates the relationship between the supply chain and the network of relationships overlaying it. The term supply chain management tends to be used to refer to management processes as well as structures of organisations.

Harland (1996) classifies SCM into four categories of use:

- *Internal Supply Chain* – this view of SCM owes a great deal to the work of Porter, (1985) on value chains and is concerned with an intra-firm approach to supply chains that involves the management of materials.
- *Dyadic relationships with immediate suppliers.*
- *The management of a chain of businesses with which you have no direct contractual relationship* (suppliers' suppliers and a customer's customer, for example).
- *The management of a network of interconnected businesses involved in the ultimate provision of a product . . . (to) end customers* (Harland, 1996).

Cox and Townsend (1998) cite the activities of Gazeley Properties Ltd as a good example of partnering and SCM. Gazeley, we are informed, '. . . attempts to manage the development supply chain in such a way as to maximise its margin while satisfying its clients' aspirations in terms of utility and cost'. If we replace the words *development supply chain* with *project*, we have a description of what all developers must be doing to remain competitive and satisfy their clients. It is, however, recognised that there is an implication that by using SCM on a construction project we are doing something more complex than managing a group of subcontractors and suppliers.

The relevance of SCM to construction (Pryke, 2001) lies not in the *existence* of supply chains, but in their *exploitation*. The management of a supply chain by a developer or contractor, implies the management of actors far removed from the dyadic contractual relationships inherent in construction contracts. Traditional (pre PPC 2000) forms of contract are based on the premise that, as an actor, one is in a relationship with another actor that instructs, pays, has control of a range of performance incentives and therefore manages ones activities. Each actor therefore is managed by the actor above in the supply chain, and in turn manages the actor or tier of actors below. Exploiting the supply chain involves communication with other actors that have been artificially separated from us by inhibiting contractual conditions. This leads us towards the concept of centrality and SCM. In order to successfully manage any supply chain we need a single actor with the authority to deal with all actors within the supply chain.

Cox and Townsend (1998) distinguish the system used by Gazeley Properties (using SCM) and those used by other, more traditional approaches, in the following terms:

- Separation of roles between end-user and fund provider and balancing the needs of these two actors;
- Use of project managers as interface with consultants and contractors;
- Concept design carried out in close consultation with end-users;
- Detailed design *may* involve input from key suppliers;
- Early participation of main contractors in design.

If we paraphrase slightly Christopher's definition (see above) we have:

The management of . . . relationships with suppliers . . . and customer to achieve greater customer value at less cost.

It is argued that this management process becomes supply chain management when it is carried out within a partnering context.

Stevens (1989), offers us a model of the transition of the firm from *stand-alone* organisation to *supply chain partners*. The four stages are as follows:

- *Baseline organisation* – classical management; motivation by profit maximisation; functional specialisation; slow to adapt to market and slow to exploit innovative opportunities.
- *Functionally integrated company* – starting to focus on customer service; competitive advantage achieved through some internal integration of disparate functions.
- *Internally integrated company* – systems approach to customer service; optimal information flow between departments; medium-term planning; cross-functional management – product focused structure.
- *Externally integrated company* – transparent system of materials and information exchange internally and externally; long-term planning and long-term relationships with partners; use of internal cross-functional management structures, product related; supplier networking groups implemented (Stevens, 1989).

Much of the literature dealing with the subject of SCM, including Stevens (1989), is not related to construction. Relating the four categories of transition to the current construction industry is disconcerting. It is argued that the vast majority of the industry falls firmly into the *baseline* category. Even those construction organisations where SCM is firmly on the agenda, show only very limited integration of disparate internal functions. In particular, cross-functional management within the organisation and the use of supplier networking groups, are particularly difficult to observe.

1.3.4   'Bottom–up' design

One of the most important changes that the construction industry must deal with in its evolution into SCM organisations is the recognition of the most appropriate location of specialist knowledge in a number of fields. Applying the principles of lean production to construction must move the location of the leadership in design from the relevant consultant to the most appropriate subcontractor, supplier or group of same. The CRINE report (http://www.crine-network.com/) borne out of the need of the North Sea Oil industry to drastically reduce its costs in the face of plummeting world oil prices, identified some important principles, which many have sought to apply to the UK construction industry. These principles were, in summary:

- Use of performance specifications to communicate interpretation of client's brief by consultant to subcontractor or supplier;
- Standard forms of contract to emphasise mutuality rather than adversarial positions;
- Use of incentives to deal more fairly with risk allocation within these non-adversarial alliances;
- Simplification of the tendering protocol and the documentation with which it is associated.

SCM articulates a process of design and financial management, the need for which must always have been present. But management of any process or system requires some focal point from which the manager can operate. The division of labour within the construction industry has meant previously that management of the whole process has been fragmented. Design, site production and component manufacture have each been managed separately. The management of these sectors have been poorly coordinated and this is partly because the conditions of contract have traditionally distinguished and separated these sectors. This tends to point to a growing need for one actor to manage the whole design/site production/component manufacture process. In terms of capacity and authority, this actor would need to be either the client or the contractor.

SCM introduces a fundamental shift in focus of responsibility and authority within the overall network of project roles. This system of evolving project roles sits within a context of competing and perhaps conflicting governance patterns. A dynamic exists between formal, contractual relationships (which initially define roles and relationships) and the less structured and formalised project management policies, such as partnering and work clusters (which both ultimately shape project roles and the way in which they are connected). These managerial approaches have a fundamental affect on actor roles and the nature and patterns of interactions between these roles.

There is a plethora of material exploring the importance and application of project management to construction projects. Increasingly, there is emphasis upon managing *programmes* of projects. Perhaps the emphasis must now change from the management of projects and programmes to the management of standing supply chains. Many would argue that integration of process, innovation and radical change in cost and value are only possible through a focus upon the activities of the supply chain. The contributions to this book have been chosen to provide a theoretical framework and case study material to explore these issues.

1.4   Overview of the Book

The text is presented in two sections; the first (Part A) deals with the establishment of a theoretical framework, which conceptualises SCM. The second part of the book deals with some important examples of the application of SCM in UK construction. Part B moves from the innovative SCM strategies adopted by two of the UK's largest clients through main contractors' issues and, finally, to an innovative approach to SCM that will appeal to the contractor and subcontractor alike.

Morledge *et al.* get things started with their chapter dealing with the concept and development of SCM in construction. The chapter traces the development of SCM in construction from its origins in manufacturing. Reference is made to the importance of both the Latham (1994) and Egan (1998) reports and the need for cultural change within the industry is emphasised; the theme of culture in supply chains is also dealt with later by Richard Fellows. Morledge *et al.* attribute the evolution of SCM to the pressure upon the construction industry for reform and the subsequent search for alternative, perhaps innovative methods and systems. The chapter ultimately questions whether or not it is feasible to implement SCM principles in construction. The transient nature of construction and its teams, especially for the one-off or low volume client, is cited as a factor. The arguments contrast with the discussion of the experiences and attributes of the very large client organisation later in the book.

The reference to culture in supply chains is built upon in a chapter by Richard Fellows. This chapter starts with an introduction to the concept of culture and asks how we fit SCM to culture. Fellows draws conceptually upon the work of Mullins (2002), providing a good fit to project 'atmosphere'. Culture values and ethics are looked at with a team-centred focus and integration is discussed. The importance of power and leverage in supply chains is emphasised. Fellows discusses the concept of cognitive dissonance and one wonders whether this helps to explain why project actors frequently express the view that 'everyone else is out of step'; frustration leading to disillusionment, then concern, followed by opportunistic behaviour. All of this is located within the context of a coalition of project actors, largely governed anonymously, and in a confrontational and formal manner, through the administration of standard forms of contracts and their written instructions. Perhaps cognitive dissonance also helps to explain why multi-disciplinary teams of actors perform badly and clients are frequently frustrated. Finally, Fellows deals with the potentially contentious (in the context of supply chains) issue of fragmentation. Fragmentation is cited as a problem. This contrasts with the work of Rimmer, where successful SCM can function hand in hand with *increased* fragmentation.

Bresnen continues the theme of culture, albeit in a rather less prominent role. Bresnen deals, in Chapter 4, with learning and innovation through collaborative relationships. After defining SCM, he considers the importance of integration in supply chains and asserts that learning in supply chains is an issue that frequently gets forgotten. Bresnen cites cultural differences as a barrier to learning in supply chains and refers to the frequently cited problem of collaborative relationships rarely cascading below the first tier suppliers in the supply chain. It is proposed that power in supply chains is based upon both leverage and expertise. Bresnen posits that there is a mismatch between systems and routines in project environments and laments the underdeveloped state of supply chain management in construction. This is a theme that Edkins links to in Chapter 9, the final chapter in the first half of the book. The conflict between transactional exchanges and project based supply chains is discussed and the importance of social networks in flows of knowledge and learning identified. Bresnen posits that there is a mismatch between systems and routines in project environments and expresses concern about the underdeveloped state of supply chain management in construction.

Skitmore and Smyth suggest that good practice in SCM should be effective in adding product and service value and this cannot be achieved without addressing marketing and pricing strategies over the long term. There is a call for the reappraisal of the way that SCM is understood. There is some criticism of the essentially deterministic approach to the discussion and analysis of SCM. Whether or not the operation and effectiveness of SCM is really influenced by custom and practice in the construction industry is an interesting point that the reader is left to consider when reading Skitmore and Smyth's chapter. The chapter starts with the recognition that SCM can take a variety of forms and is perceived differently within different industries. Marketing theory is suggested as an alternative to what usually comprises a procurement-driven approach to projects. Pricing theory is discussed and the predominance of the price-dominated marketing mix in construction is questioned. It is suggested that applying SCM within a risk minimisation and transaction context leads to cost cutting or the reduction of added value potential and continuous improvement is jeopardised. Skitmore and Smyth provide a suitably thought provoking end to Part A of the book.

Rimmer's chapter opens the second (applied) part of this book. The author has made a major contribution to the reform of the construction industry over the last two decades. Rimmer provides case study material relating to the organisation that he helped manage and through which he was able to implement many important SCM initiatives. The chapter deals with the important role that the hands-on clients have to play in the management of the construction supply chain. Uniquely, the importance of SCM during the period over which the client must finance the site purchase and organise design and construction prior to financial payback is high-lighted. Rimmer provides an overview of each of the types of procurement and provides a graphical model showing three different types of link between supply chain firms – *direct* (involving payment); *management relationships*; and *design relationships* (both the latter involving formal instructions). Rimmer identifies the problem that prevails in construction whereby the professionals or consultants are typically trusted and the contractors and subcontractors are mistrusted; this is cited as an important barrier to the effective implementation of SCM. The importance of collaborative relation-ships in construction project teams is emphasised.

Potts takes on the task of documenting and discussing the high profile supply chain managing efforts of British Airports Authority (BAA). He dis-cusses how the client for the Heathrow Express at the time of a highly publicised tunnel collapse during construction, used SCM principles to recover from the disaster and enable the project team to function effectively and complete the project efficiently. The Genesis project provided a pilot study for the techniques and systems proposed in the management of BAA's Terminal 5 at Heathrow. Potts relates the way in which BAA investigated the cause of failure in 'mega-projects' and found that the way in which the supply chain was engaged and risk management systems used were impor-tant factors. BAA's development of supply chain strategies and systems evolved through the exploitation of three key initiatives – partnering; inno-vation and best practice; and the framework agreement programme. The framework agreement programme, BAA's structured partnering arrange-ment, provided a comprehensive, if somewhat bureaucratic model for others to emulate. Potts cites the main motivations for the SCM initiatives at BAA as Jones *et al.*'s famous (non-construction) book, *The Machine that Changed the World* (1990) and the Egan (1998) report. The chapter finally looks at the detail of BAA Heathrow Terminal 5 (T5): BAA envisaged the virtual firm, where the principles of partnering were applied and developed to their full. The contract documents comprised the T5 Agreement, which in simple terms moved away from the traditional dyadic contract forms epitomised by the Joint Contracts Tribunal (JCT) forms of contract: the documents effectively map the extent of the project network and the nature of the relationships between the various project actors. A number of T5 best practice principles are set out bringing this review of 15 years of BAA's development to a close.

In King and Pitt's chapter we move very firmly from the clients' perspec-tive on SCM over to the main contractor's view. This chapter is a collabora-tive venture between the academic and the practitioner and essentially deals with Pitt's experience of SCM with Bluestone plc, which was bought out by Morgan Ashurst during the writing of the chapter. The chapter differentiates between organisational and project supply chains and essentially challenges the main premise of the two previous chapters in terms of which project actor is best placed to manage the supply chain. The research on which the chapter was based used an innovative research method – action research;

this research involved five dominant themes comprising relationships, culture, consolidation, consistency and cost. The chapter explores Green's (1999) dismissal of cultural change and challenges his questioning of the transferability of good practice in SCM into construction. The chapter also draws on Cox and Ireland's (2002) important work in the area of leverage in supply chains and the effect and location of power. King and Pitt regret the absence of the equivalent to prime contracting (see Holti *et al.*, 2000) in the private sector and conclude, perhaps uniquely amongst the chapters presented here, that SCM does not find an immediately successful application in construction because the quality of project relationships are of more overwhelming importance, regardless of whether or not SCM principles are implemented.

Power is a theme that Edkins considers. The conflict between transactional exchanges and project-based supply chains is discussed, and the importance of social networks in flows of knowledge and learning identified. Edkins starts this chapter on uncertainty management by considering the totally vertically integrated organisation (like Ford cars in its early days), where much of the final product is created from raw materials by the final assembler of the product. Edkins refers to the related issues of risk transfer and leverage in the supply chain and whether the construction industry should look to contractual remedies or adopt a relationship management approach. The mapping of risk and certainty with benefit and threat are considered. Edkins draws upon some important case study material, including the 1987 Space Shuttle Challenger disaster. The place that is occupied by construction alongside other types of project–based industry is discussed. The role that contracts have to play in the establishment of expectations, roles, responsibilities, incentives and penalties is covered and the chapter concludes with the proposition that risk is allocated to the firm in the supply chain which has the least leverage and is the least able to defend itself.

Smyth's chapter on franchising in construction finishes Part B. This thought provoking chapter leaves the reader wondering whether the concept of franchising is an application of SCM or simply *an alternative* means to structure and control supply chains. Smyth identifies one of the perennial problems facing the construction project manager of trying to control a number of organisations that are not owned by their firm. Smyth identifies a model for franchising in construction that involves the use of standard products and processes, centrally provided marketing and advertising; and some investment by the franchisee. The franchiser provides a brand, quality control, codes of behaviour and some of the administrative support needed to gain workload. The franchisee, whether a large subcontractor or a smaller firm, gains the benefits of being associated with and within a large branded operation but retains control of the franchised business unit and gains through knowledge transfer. Smyth identified two classifications of franchising: management franchising and 'man and van' franchising, the latter as a group gaining more from franchising than the former. Franchising is put forward as an alternative to centralised SCM in construction, based upon the premise that there are similar benefits to the project that franchising can achieve. Importantly, franchising also has the benefit of reaching the lower tiers of the supply chain, and finally that franchising has the potential to lift the image of construction through a renewed focus on customer care, brand image and the uniform service.

1.5   Summary

This first chapter began with the proposition that the term SCM is, perhaps, inaccurate. It was argued that SCM exists at a high level of abstraction and to observe the operation of the supply chain we might want to explore the network of relationships that comprise the supply chain. None the less, the terms supply chain and supply chain management are in common use and provide the key terms for this text. The motivations behind the pursuit of SCM and the context in which SCM evolved are explored.

An overview of the chapters in the book was provided and the clear definition between the conceptual material of Part A and the applied and case-study material of Part B is made. There is a brief summary of the historical development of both partnering and SCM to provide a context for the discussion which follows.

References

Ayers, J.B. (2004) *Supply Chain Project Management: a Structured Collaborative and Measurable Approach*. CRC Press LLC, Florida, USA.

[Banwell Report] Ministry of Public Building and Works (1964) *The Placing and Management of Contracts for Building and Civil Engineering Works*. HMSO, London.

Barlow, J., Cohen, M., Jashpara, A. and Simpson, Y. (1997) *Partnering: Revealing the Realities in the Construction Industry*. The Policy Press, Bristol.

Betts, M. and Wood-Harper, T. (1994) Re-engineering construction: a new management research agenda. *Construction Management and Economics*, 12(6), Nov, 551–556.

Bresnen, M. (1997) *Constructing Partnerships: Re-Engineering the Knowledge Base of* Contracting. Presented at a Construction Economics Group seminar, University of Westminster, 29/5/97.

Bresnen, M. (2000) Partnering in construction: a critical review of issues, problems and dilemmas. *Construction Management and Economics*, 18, 229–237.

Cain, T.C. (2003) *Building Down Barriers: A Guide to Construction Best Practice*. Taylor and Francis, London.

Cherns, A.B. and Bryant, D.T. (1984) Studying the client's role in project *management*. *Construction Management and Economics*, 1, 177–184.

[CBPP] Construction Best Practice Programme (2003) *Fact Sheet on Supply Chain Management*. Construction Best Practice Programme, Garston.

Christopher, M. (1997) *Marketing Logistics*. Butterworth Heinemann, Oxford.

Copacino, W.C. (1997) *Supply Chain Management: The Basics and Beyond*. St Lucie Press, Boca Raton, Florida, USA.

Cox, A. and Ireland, P. (2002) Managing construction supply chains: a common sense approach. *Engineering, Construction and Architectural Management*, 9(5), 409–418.

Cox, A. and Townsend, M. (1998), *Strategic Procurement in Construction: Towards Better Practice in the Management of Construction Supply Chains*. Thomas Telford Publishing, London.

[Egan report] DETR (1998) *Rethinking Construction: The Report of the Construction Task Force to the Deputy Prime Minister, John Prescott, on the Scope for Improving the Quality and Efficiencies of UK construction*. DETR at www.construction.detr.gov.uk/vis/rethink.

Dyer, J.H. (1996) *How Chrysler Created an American Keiretsu*. Harvard Business Review July-August 1996, USA.

Gattorna, J.L. and Walters, D.W. (1996) *Managing the Supply Chain: A Strategic Perspective*, Macmillan Business, London.

Green, S.J. (1999) The Missing Arguments of Lean Construction, *Construction Management and Economics*, 17, 13–137.

Green, S.J. (2006) Discourse and Fashion in Supply Chain Management. *In* Pryke, S.D. and Smyth, H.S. (2006) *The Management of Projects: A Relationship Management Approach*. Blackwell, Oxford.

Green, S.D. and May, S.C. (2003) Re-engineering construction: going against the grain. *Building Research & Information*, 31(2), 97–106.

Gummesson, E. (2001) *Total Relationship Marketing*. Butterworth-Heinemann, Oxford.

Harland, C.M. (1996) Supply chain management: relationships, chains and networks. *British Journal of Management*, Vol. 7, Special issue, S63–S80, March 1996.

Harlow, P. [ed] (1994) *Mythology and Reality: The Perpetuation of Mistrust in the Building Industry*. Construction Papers, CIOB.

Higgin, G. and Jessop, N. (1965) *Communications in the Building Industry: The Report of a Pilot Study*. Tavistock Publications, London.

Holti, R., Nicolini, D. and Smalley, M. (2000), *Prime Contractor's Handbook of Supply Chain Management*. Tavistock Institute, London.

Howell, G., Miles, R., Fehlig, C. and Ballard, G. (1996) *Beyond Partnering: Toward a New Approach to Project Management*, Proceedings of Conference on Alternative Dispute Resolution sponsored by the Construction Industry Institute and the University of Texas Law School, San Antonio, Texas, April 1996.

Koskela, L. (1992) *Application of the New Production Philosophy to Construction, Technical Report 72*. Center for Integrated Facility Management, Stanford University, Stanford.

Koskela, L. (2000) *An Exploration towards a Production Theory and its Application to Construction*, Report 408, VTT, Espoo.

La Londe, B. (2003) Five principles of supply chain management. *Supply Chain Management*, 7(3), 7–8.

Latham, Sir M. (1994) *Constructing the Team: Joint Review of Procurement and Contractual Arrangements in the United Kingdom*. HMSO, London.

London, K. and Kenley, R. (2001) An industrial organization economic supply chain approach for the construction industry: a review. *Construction Economics and Management*, 19, 777–788.

Loraine, R.K. (1994) Project specific partnering. *Engineering, Construction and Architectural Management*, 1, 5–16.

Morris, P.W.G. (1994) *The Management of Projects*, Thomas Telford, London.

Morris, P.W.G. and Pinto, J.K. (eds) (2004) *The Wiley Guide to Managing Projects*, John Wiley & Sons, New York.

Mullins, L.J. (2002) *Management and Organisational Behaviour (6 edn.)*. Harlow: Prentice Hall.

New, S. and Westbrook, R. [Eds.] (2004) *Understanding Supply Chains: Concepts, Critiques and Futures*. Oxford University Press, Oxford.

Porter, M.E. (1985) *Competitive Advantage*. Free Press, New York.

Project Management Institute (2000) *A Guide to the Project Management Body of Knowledge*. Project Management Institute, Pennsylvania, USA.

Pryke, S.D. (2001) *UK Construction in Transition: Developing a Social Network Approach to the Evaluation of New Procurement and Management Strategies*. PhD in Building Management, The Bartlett School, UCL.

Pryke, S.D. and Smyth, H.S. (2006) *The Management of Complex Projects: A Relationship Management Approach*. Blackwell, Oxford.

Ptak, C.A. and Schragenheim, E. (2000) *ERP: Tools, Techniques and Applications for Integrating the Supply Chain*. St. Lucie Press, Boca Raton, Florida, USA.

[Simon] The Central Council for Works and Buildings (1944) *The Placing and Management of Building Contracts*. HMSO, London.

Simon, H.A. (1976) [3rd. edn] *Administrative Behavior: A Study of Decision-making Processes in Administrative Organisations*, Free Press, New York.

Smyth, H.J. (2006) *Competition, Commercial Management of Projects: Defining the Discipline*. Lowe, D. and Leiringer, R. (eds.). Blackwell, Oxford.

Stevens, G. (1989) Integrating the supply chain. *International Journal of Physical Distribution and Materials Management*, 9(8), 3–8.

Turner J.R. (1999) *The Handbook of Project-based Management: Improving the Processes for Achieving Strategic Objectives* (2nd ed.). McGraw-Hill, Maidenhead.

Turner, J.R. and Mûller, R. (2003) On the nature of the project as a temporary organization. *International Journal of Project Management*, 21(1), 1–8.

Virhoef, R. and Koskela, L. (2000) The four roles of supply chain management in construction. *European Journal of Purchasing and Supply Management*, 6, 169–178.

Wenger, E. (1998) *Communities of Practice: Learning, Meaning, and Identity*. Cambridge University Press, Cambridge.

Winch, G. (1989) The construction firm and the construction project: a ransaction cost approach. *Construction Management and Economics*, 7, 331–345.

Winch, G. (2000) Institutional reform in British construction: partnering and private finance. *Building Research and Information*, 28(2), 141–155.

Womack, J.P., Jones, D.T. and Roos, D. (1990) *The Machine that Changed the World*, Maxwell Macmillan International, Oxford.

Winch, G.M. (2002) *Managing the Construction Project*. Blackwell, Oxford.

Winch, G.M. (2004) Managing project stakeholders. In: *The Wiley Guide to Managing Projects*, eds. P W G Morris and J K Pinto, John Wiley & Sons, New York.

Part A
Concepts

The first four chapters are written by leading international academics with the purpose of providing a framework for the discussion of supply chain management (SCM). An introduction to the main concepts and some history of the development of SCM contributed by Roy Morledge and his colleagues provides context for Richard Fellows' important chapter on culture in supply chains. Mike Bresnen looks at knowledge, learning and innovation in supply chains and links to the previous chapter through the discussion of cultural issues in learning. Part A is rounded off by Martin Skitmore and Hedley Smyth who provide a discussion on the relevance of marketing theory and pricing strategy in supply chains and raise the issue of a value added approach to competition in construction.

2

The Concept and Development of Supply Chain Management in the UK Construction Industry

Roy Morledge, Andrew Knight and Mohieddin Grada

2.1   Introduction

Supply Chain Management (SCM) techniques have been successfully used in various industries such as food and manufacturing for decades. The supply chain in these industries encompasses all the activities associated with processing: from raw materials to completion of the end product. This includes procurement, production scheduling, order processing, inventory management, transport, storage, customer service and all the necessary supporting information systems. It is usually an ongoing process focused upon specific products, which are repeatedly manufactured or purchased. Its management consists of a stable group of interacting partners with a mutual interest in improving product quality and process efficiency.

SCM strategies, as they are adopted in the manufacturing industries, assume an ongoing process where supplier and customer experience involves frequent transactions for the same or similar products. They are seen as a key to maintaining quality and facilitating innovation and measurable improvement. To a large extent, their success depends upon a long-term shared culture, both within and outside a particular organisation.

To use the term 'Supply Chain Management' in the context of the current UK construction industry suggests that it is possible to adopt those practices, which have proved to be successful elsewhere, without significantly adapting them to reflect the particular nature of the industry and its culture. Hence, the aim of this chapter is to broadly contextualise and explore the problem of SCM in the construction industry; therefore, it will complement other chapters focused on more specific case studies and areas elsewhere within this book. First, the particular characteristics of the UK construction industry are explored. Second, the history of some key initiatives is introduced with particular reference to SCM issues. Third, the historical development of SCM is discussed with particular reference to earlier approaches such as logistics. The concept of SCM is then analysed through the examination of various definitions to find a common understanding of the term. Fourth, with reference back to the characteristics of the construction industry, the application of SCM is discussed along with the challenges this presents.

Our aim is to provide the reader with a broad recent historical overview and as such we refer to two major reports throughout: Latham (1994) and Egan (1998). Although the criticisms of the construction industry outlined in these reports may appear to be dated by some, these reports are important as they acted as a catalyst for change. Additionally, many of the cultural problems highlighted in the industry may have improved over the last decade, but they still cause problems today. Moreover, whilst many leading clients and contractors may have improved their procurement systems, many others still operate using what might be regarded as traditional systems. Hence, it is essential to recognise the slow nature of cultural change to appreciate the true challenges still facing many parts of the industry.

2.2   Characteristics of Construction Industry

The construction industry in the UK exhibits characteristics that differentiate it from other industrial sectors. It is one of the UK's key sectors contributing 8.2% of Gross Value Added and employing 2.1 million people in 250,000 firms (BERR 2007). The UK Standard Industrial Classification (UK SIC 2003, p. 117) states that construction includes general construction and special trade construction for buildings and civil engineering, building installation and building completion. It includes new work, repairs, additions and alterations, the erection of prefabricated buildings or structures on the site, and constructions of a temporary nature.

The UK construction industry has been accused of being, at its worst, wasteful, inefficient and ineffective. There are some concerns that the industry is not performing to its full potential. These concerns mainly focus on areas regarding the industry's profitability margins, its clients' satisfaction and the fragmentation of the construction procurement process. Competitive pressures from within the industry, as well as external political, economic and other considerations are forcing the industry to re-examine and improve its performance (Anumba *et al.* 2000). An analysis of the key characteristics of the construction industry indicates that the problems facing construction can be categorised into the following five broad areas.

2.2.1   Fragmentation

The construction industry has long been recognised as having problems in its structure, particularly with fragmentation, which has resulted in poor performance (Latham, 1994; Egan, 1998). Uniqueness (see also below), immobility and variety are three distinctive features of construction output that flow from the fragmentation in construction. These features, it is argued, are factors in the tendency of the construction industry towards low productivity, poor value for money and mediocre overall client satisfaction (Latham, 1994), especially when compared with other industry sectors.

As a consequence of the uncertainty for the main contractor in obtaining continuous work, with the need to accommodate the different features and requirements of each project, subcontracting has been adopted as the dominant approach (Cox and Townsend 1998, p. 21), which by its very nature, results in further fragmentation.

2.2.2   Adversarial relationships

The construction supply chain has become increasingly fragmented for the reasons outlined above. Increased fragmentation brings increased transaction volumes at lower average values and inevitably higher levels of opportunism, particularly in the context of low barriers to entry. The industry had become less trusting, more self-interested and adversarial. The adversarial attitude of the UK construction industry has been recognised problem for many years (Cox and Townsend 1998, p. 29). Performance and innovation in construction are significantly hindered by adversarial relationships and fragmented processes. In order to minimise their own exposure to risk, each party in the supply chain attempts to extract maximum reward for minimum risk that is normally achieved by means of non-legitimate risk transfer (passing risk down to the next level in the supply chain). This way of thinking has resulted in an industry structure with various interfaces, which are points of tension and conflict, which eventually leads to increased cost and reduced efficiency (Cox and Townsend 1998, p. 31). This scenario is illustrated in Figure 2.1.

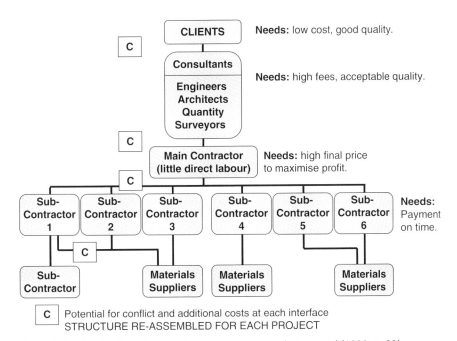

Figure 2.1   Existing industry structure. Source: Cox and Townsend (1998, p. 33).

2.2.3   Project uniqueness

The construction industry (with the possible exception of some responsive repairs) is a project-based industry. The characteristics of a specific project, and hence its degree of uniqueness, is determined by a number of factors. Consequently, the assessment of these project features determines the resources needed for a project, and selection of the most appropriate supply chains needed to deliver clusters (Gray, 1996) of resources and services for the project as a whole. This diversity and uniqueness means that construction projects are very often 'bespoke' as the requirements and specifications of technologies for specific clients determine their characteristics. Projects involve assembling materials and components designed and produced by a multitude of suppliers, working in a diversity of disciplines and technologies in order to produce a product for particular client. This diversity of product technologies, which has to be reorganised with each new construction project, coupled with discontinuous demand from a large percentage of construction's clients, accounts for the transient nature of the relationships between the demand and supply side of the industry. In addition, with the increasing shift from on-site to off-site production, managing construction projects involves integrating diverse and complex supply systems in which a growing amount of value of the product is added (Jones and Saad 2003, p. 12).

2.3.4   Separation of design and production

One of the main problems in construction is the extent to which the industry separates design from production. This particular characteristic of the industry is still common in spite of the deficiencies of traditional procurement and the benefits offered by newer and more flexible approaches (Cooke and Williams 2004, p. 2) The separation of the design and production process in the construction industry, particularly in the building sector, and the consequent difficulties that can arise during construction projects, has been the subject of wide criticism by a number of industry reports such as Latham (1994) and Egan (1998). Consequently, there have been many calls to bridge this gap by creating a seamless supply chain whereby the interface between various phases of the project's life cycle are integrated with one another. It is an anomaly that design and production are commonly separated at the highest tier of the supply chain (main contractor) but commonly integrated in the tiers below this.

2.2.5   Competitive tendering

In most countries, construction companies are selected to undertake construction projects and the price for their work is established by competition (Griffith *et al.* 2003). Unlike manufacturing, construction projects are not priced and advertised for sale (manufactured speculatively, without prior orders from customers), but instead uniquely priced after a negotiation or bidding process. Since the mid 1990s joint government and construction industry initiatives (Latham 1994; Egan 1998) have encouraged construction clients to adopt different strategies to procure work. Whilst regular, experienced and informed construction clients have begun to adopt alternative procurement strategies, there is little evidence to show that the majority of inexperienced irregular purchasers have done so. By far the dominant strategy adopted is the traditional design-bid-build approach with the lowest bidder winning the work (RICS 2006). There is no doubt that competition used in this way serves to drive down prices.

Adopting 'low bid wins' strategies results in a number of undesirable outcomes, particularly where the design is already established. These are:

• production processes that are geared to lowest cost rather than to 'right first time' or to 'best value';
• bidding processes that encourage opportunism - where suppliers will agree to almost any conditions and requirements to get the work and attempt to improve profit levels on the project through reductions in quality of materials, or the negotiation of disproportionately high rates for variation works;
• an inability and unwillingness to cooperate in specialist design, innovation or collaborative problem solving.

'Low bid wins' procurement has been blamed for, amongst others, late completion, overspends on client budgets and product / workmanship. Alternatives do exist, but require attitude change within the construction sector and its professions.

2.3   Government Initiatives in the Construction Industry

There has been a succession of reports into the state of the UK construction industry and there have been many calls for action to improve its performance and competitiveness. Latham (1994) and Egan (1998) are two of the most recent and influential Government funded reports. These initiatives in the construction sector have highlighted the need to spread good practice within the sector and to transfer good practice from others. This has seen the development of the Rethinking Construction Initiative, which has promoted the Movement for Innovation and the Construction Best Practice Programme, now known as Constructing Excellence. One of the central tenets of Rethinking Construction is to learn about strategies, methods and approaches that have proved successful in particular companies and that have a technology transfer opportunity for the construction sectors.

Latham (1994) recommended that alternative arrangements for contractual relationships were an essential element in creating improvement in construction. By so doing, Latham asserted that it should be possible to attempt to achieve in the order of thirty per cent improvement in productivity. The Egan Report, whilst being entirely sympathetic to Latham's recommendations, proposed that much more needed to be done in order to achieve the sort of radical improvement that many have suggested is possible in construction.

Egan (1998, p. 3) stated that:

> *'At its best the UK construction industry displays excellence. But there is no doubt that substantial improvements in quality and efficiency are possible.'*

The report suggested that, if the construction industry is to create conditions favourable to radical improvement, it must make certain changes, the most notable of which were:

- Modernisation;
- Increased spending on training, research and development;
- Creation of better relationships between contractors and clients;
- Increase of the use of standardisation and pre-assembly;
- Application of performance tools and techniques.

In comparison with other sectors, the construction industry is relatively unsophisticated in its approach to the supply chain. The Egan report argues that construction can learn from the experiences of other industries. Specifically, the report refers to manufacturing and service industries in which the authors assert that there have been 'increases in efficiency and transformation of companies which a decade ago nobody would have believed possible' (Egan 1989, p. 11). Even though the precise effect that these reports have had in changing the UK construction industry is debatable, one thing that all these reports tend to agree upon is that the resources used by the construction industry can be made to perform more effectively and that as a direct result all parties could enjoy greater benefit. According to Egan (1998):

> *'Construction businesses are beginning to realize that their success is increasingly dependent on the organizations they supply to and buy from, and that for continued success they need to cooperate and collaborate across customer/supplier interfaces.'*

The need for UK construction companies to become more efficient has resulted in an interest in various management systems as a means to achieve the recommendations set out in these reports. Both reports have concentrated attention on the need for the construction industry to improve the efficiency with which it operates its supply network. Currently, SCM provides an important focus or perhaps conceptual framework for construction project management research, in recognition of the interface between organisations as a source of 'friction' and a potential source of improvement in many aspects of the construction and design process. Having outlined the key features of the construction industry that have shaped its ability to innovate, change and respond to new challenges, the next section attempts to review the development of the concept of SCM.

2.4   The Historical Development of Supply Chain Management

SCM, as a term, first appeared in the early 1980s to describe:

> '...the range of activities co-ordinated by an organisation to procure and manage.'

SCM is a concept that originated and developed in the manufacturing industry. The first signs of SCM were perceptible in the Just in Time (JIT) delivery system as part of the Toyota Production System (Shingo 1988). Harland *et al.* (1999) indicate that the evolution of SCM theory is driven by rapid changes in global business practice. They contend that the worldwide recession of the late 1980s and early 1990s forced companies to re-examine, at a strategic level, the ways in which they aimed to add value and reduce costs throughout their business. Initially, the term referred to an internal focus bounded by a single organisation and how that organisation sourced and procured supplies, managed their internal inventory and moved goods onto its customers. It was recognised that this understanding was inadequate and that the reality of managing supplies meant that supply chains extended beyond the purchasing organisation and into its successive lower tiers (suppliers and their suppliers' suppliers) [Christopher 2005, p. 5].

SCM, as an area of study, is a recent phenomenon and yet is clearly related to logistics. It is a common notion that logistics involves the movement of physical goods from one location to another. As long ago as the construction of the great pyramids, man was concerned with how to move materials to a construction site. Human migration from Europe to the Americas is another example of significant logistical challenges. The term logistics was originally first used in the military environment. The study of logistics received much attention from the armed forces during both World Wars. The Second World War necessitated greater movement of troops and supplies than any other period in history; logistics proved a crucial factor in its outcome and indeed the success or failure of many military conflicts. Following the war, logistical concepts were given more attention in the business world (Christopher 2005, p. 3; Long 2004, p. 4). Figure 2.2 illustrates the evolution of logistics as a discipline, into SCM.

There are many ways of defining logistics, but the Council of Supply Chain Management Professionals offers a useful standard:

> 'The part of the supply chain that plans, implements, and controls the efficient, effective flow and storage of goods, services and related information from the point of origin to the point of consumption in order to meet customers' requirements.'

According to Long (2004, p. 10), from the point of view of a company, there are three distinct areas to logistics. *Inbound logistics* includes sourcing and materials management; *operations logistics*, closely related to material management emphasizing the way logistics affects operations; and finally *outbound logistics*, also known as physical distribution, refers to the way

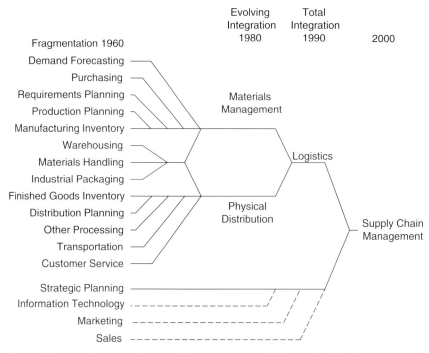

Figure 2.2 Logistics evolution to SCM. Source: Long (2004, p. 6). With kind permission of Springer Science and Business Media.

the product is physically delivered to customers. This distinction is vital in determining the influence or power of each party in the supply chain.

The terms 'SCM' and 'logistics' are often confused and viewed as overlapping, depending on the definition used by an organization. According to Lummus *et al.* (2001), there is confusion and disagreement among general business practitioners and operations professionals regarding the terms *logistics* and *SCM*. Various formal definitions have been offered for both terms. With the increased interest in SCM, several authors have discussed the differences between this newer term (SCM) and logistics. Cooper *et al.* (1997) stated that an understanding of SCM is not appreciably different from the understanding of integrated logistics management. However, logistics can be thought of as a planning orientation and framework that seeks to create a single plan for the flow of product and information through a business. SCM builds upon this framework and seeks to achieve linkage and co-ordination between the processes of other entities in the pipeline, that is suppliers and customers and client organisation itself (Christopher 2005, p. 4).

2.5   The Concept of Supply Chain Management

It can be argued that SCM is not just another name for logistics. SCM goes further and includes elements that are not typically included in a definition of logistics, such as information systems as well as the integration and coordination of planning and control activities. As logistics primarily deals with the flows to, in and out of companies, with an *intra*-organizational perspective, SCM is a development that deals with the *inter*-organizational view of logistics alongside the intra-organisational perspective.

Various definitions of a supply chain have been offered in the past as the concept has gained popularity. Many definitions describe SCM as the chain linking each element of the manufacturing and supply process from raw materials to end users, encompassing several organisational boundaries. This is well summarised by Christopher (1992, p. 18) who defined supply chain as:

> '*The management of upstream and downstream relationships with suppliers and customers to deliver superior customer value at less cost to the supply chain as a whole.*'

In order to reflect the fact that there will normally be multiple suppliers, and suppliers to those suppliers, as well as multiple customers, and customers' customers, to be included in the total system, Christopher (2005, p. 5) argues that the word 'chain' should be replaced by 'network'. He also argues that since the chain should be driven by the market, not by the suppliers, the phrase SCM should be termed demand chain management.

According to Christopher (1992, p. 15), a supply chain is 'the network of organisation that are involved through upstream and downstream linkages, in the different processes and activities that produce value in the form of products services in the hands of the ultimate consumer'. Moreover, Christopher defines the objective of SCM in a relative manner as delivering superior value at less cost.

In summary, key words in typical definitions of SCM include: *network, integrative, channel, upstream, downstream linkages, ultimate user* and *value*. These definitions link SCM with the integration of systems and processes within and between organisations, including the upstream suppliers and downstream customers and involving methods of reducing waste and adding value across the entire process. Porter (1985, p. 48) also emphasises the importance of effective linkages among the activities in the value chain:

> '*Linkages can lead to competitive advantage in two ways: optimization and coordination. Linkages often reflect tradeoffs among activities to achieve the same overall result . . . A firm must optimize such linkages reflecting its strategy in order to achieve competitive advantage . . . The ability to coordinate linkages often reduces costs or enhances differentiation . . . Linkages imply that a firm's cost or differentiation is not merely the result of efforts to reduce cost or improve performance in each value chain activity individually.*' (Porter, 1985:48)

The importance of effective linkages is also stressed by Jones and Saad (2003, p. 242). They state that 'the SCM has shifted the emphasis from internal structure to external linkages and processes, and is dependent on the interaction between the organisation and its external environment with strong feedback linkages and collective learning'. In the context of the construction industry, SCM can be regarded as the process of strategic management of information flow, activities, tasks and processes, involving various networks of organisations and linkages (upstream and downstream), throughout a project life cycle. In terms of the foregoing, the upstream activities within construction SCM, in relation to the position of a main contractor, consists of the activities and tasks leading to preparation for production on site, involving construction clients and design teams. The downstream consists of activities and tasks in the delivery of the construction product involving construction suppliers, subcontractors, and specialist contractors interrelating with the main contractor. In the construction industry, particularly on a larger project, which involves a significant number of separate supplying organisations, the complexity of the network can often be very significant (Briscoe *et al.* 2001).

The concept of SCM is based on the notion that supply chains rather than single business units are competing with each other (Geir *et al.* 2006). According to Christopher (2005, p. 18), leading-edge companies recognize the fallacy of simply transferring costs upstream or downstream and instead seek to make the supply chain as a whole more competitive through the value it adds and the costs that it reduces overall. In part, the concept of SCM represents a logical continuation of previous management development principles such as Total Quality Management (TQM), Business Process Redesign (BPR) and Just in Time (JIT). Van der Veen and Robben (1997) argue that SCM is combining the particular features of these three techniques.

According to Christopher (2005, p. 18), 'Leading-edge companies have realized that the real competition is not company against company, but rather supply chain against supply chain'. This is a change of system boundaries where a supply chain can be viewed as a system of companies, which in their turn can be viewed as a system of processes and functions.

Harland (1996) describes a four-stage supply chain classification, outlining four main uses for the term SCM:

- The internal supply chain integrates business functions involved in the flow of materials and information from the inbound to outbound ends of the business.
- The management of a dyadic, or two-party, relationship with immediate suppliers.
- The management of a chain of businesses including a supplier, a supplier's supplier, a customer and a customer's customer and so on.
- The management of a network of interconnected businesses involved in the ultimate provision of product and service packages required by end customers.

Croom *et al.* (2000) identified the three inter-organisational supply chain levels as dyadic, chain and network. This classification contrasts the nature of the exchange relationships against the supply chain types, suggesting that the type of exchange relationship is a fundamental building block of SCM.

In summary, the literature has identified that SCM entails something more than logistics management. It concerns different levels of analysis – dyadic, chain and network – and involves the exchange of assets, information and knowledge between companies that are interlinked in the provision of goods or services. This process ranges from inception to final consumption and includes the management of inter-organisational relationships that are developed between these companies.

Owing to the unique characteristics of the construction industry, it has been argued that management techniques and principles, such as SCM, applied to other product producing industries, such as the manufacturing industry, often need to be modified before they are applied to the construction industry, otherwise their effectiveness is very limited. The following section will evaluate the relevance, adoption and implementation of SCM in construction. It seeks to address whether construction has the ability to use and sustain SCM between its large number of discrete work packages and across organisational boundaries. This section also highlights the fundamental key characteristics of construction, which need to be taken into account for a successful implementation of this technique.

2.6   The Application of Supply Chain Management Techniques in the Construction Industry

SCM has a critical role to play in improving overall performance in construction, but remains at a very early stage of development (Jones and Saad 2003, p. 219). However, the industry is becoming increasingly aware of the necessity to change current working practices and the attitudes they represent (Pearson 1999). A number of organisations (for example; Balfour Beatty and Tarmac) within the construction industry and their clients (for example; BAA (see also Chapter Eight), the Ministry of Defence and Tesco) have developed SCM techniques to rationalise their supplier base. Unlike in the construction industry, SCM has been practised widely for many years in other industries, particularly in the manufacturing sector. Effective SCM has helped numerous industry sectors to improve their competitiveness in an increasingly global market place. The SCM in these industries encompasses all those activities associated with processing from raw materials to completion of the end product for the client or customer. It is usually an on-going process focused upon specific products that are repeatedly manufactured or purchased. SCM consists of a stable group of interacting partners with a mutual interest in improving product quality and process efficiency. Unlike the manufacturing industry, the construction industry lacks standardisation. Through the use of standard parts and components, the manufacturing industry has been successfully generating greater levels of productivity and quality control. The construction site is effectively an *ad hoc* factory, temporarily created to manufacture a prototype product (Cox and Townsend 1998, p. 255). To what extent the construction industry can standardise its product is essential to the development of effective SCM.

The **Supply Chain Council**, an independent non-profit organisation, has developed a SCM maturity model (McCormack *et al.* 2004). The model defines the following SCM maturity levels:

- *Level 1 – Ad hoc* – The supply chain and its practices are unstructured and ill-defined. Processes, activities and organisational structures are not based on horizontal processes, while process performance is unpredictable. SCM costs are high, customer satisfaction is low, functional co-operation is also low.
- *Level 2 – Defined* – Basic SCM processes are defined and documented, but the activities and organisation basically remain traditional. SCM costs remain high, customer satisfaction has improved, but is still low.
- *Level 3 – Linked* – This level can be considered a breakthrough where cooperation between company departments, vendors and customers is established. SCM costs begin decreasing and customer satisfaction begins to show a marked improvement.
- *Level 4 – Integrated* – The company, its vendors and suppliers co-operate on the process level. Organisational structures are based on SCM procedures; SCM performance measures and management systems are applied. Advanced SCM practices, like collaborative forecasting with other members of a supply chain, form. As a consequence, SCM costs are dramatically reduced.
- *Level 5 – Extended* – Competition is based on supply chains. Collaboration between companies is on the highest level, multi-firm SCM teams with common processes, goals and broad authority form.

The five stages of maturity show the progression of activities toward effective SCM and process maturity. Each level contains characteristics associated with process maturity such as predictability, capability, control, effectiveness and efficiency. Placing the construction industry within the five levels described above, it can be argued that the UK construction industry has to some extent utilised SCM techniques for years where ad hoc supply chains of subcontractors are assembled and then disassembled at the end of each project. This basically can be attributed to the one-off nature of construction projects coupled with discontinuous demand. It can be argued that this type of traditional supply network is unlikely to maximise the value for parties involved in the supply chain.

The traditional construction project supply chain can be described as a series of sequential operations by groups of people who have limited concern about other stakeholders. Most construction projects are procured through a method by which a defined project forms the focus for a building process carried out by a contractor, who traditionally obtains the work by bidding the lowest price for carrying out the project. The appointed contractor will outsource or subcontract the majority of the work to a number of relatively small subcontractors who will usually win the work on exactly the same basis. The number of subcontractors will vary with the complexity and nature of the project. These contractors and subcontractors typically focus upon meeting their contractual requirements for the lowest possible cost. There may be limited commitment to the client's primary objectives or to any perceived project team. A parallel, but separate supply chain managed by the client, or client's project manager, will in most cases include the procurement of the financial resources to support the project and the procurement of the design process itself.

Traditional unmanaged supply chains are characterised by short-term focus, with little concern for mutual long-term success; adversarial relationships between customers and suppliers, including 'win-lose' negotiations; little regard for sharing benefits and risks; and primary emphasis on cost and delivery, with little concern for added value. As a consequence, the traditional supply chains in a construction project are complex and temporary, involving participants who may not contribute, other than to complete their small, often isolated, part of a one-off project. A team culture focused on the particular needs of the client or the project rarely exists. Different approaches to managing such a supply chain are therefore required if the potential benefits are to be achieved. The culture within which SCM can be developed may not exist in traditional procurement methods, but can be created if the value of such change can be shown to be significant.

Despite the limited value and benefits of the ad hoc supply chain structure, it has worked with varying degrees of success in construction projects, owing to the unpredictable nature of the construction process (Cartlidge, 2004: p. 127). According to Vrijhoef and Koskela (2000), there are four levels, termed 'roles of SCM', of implementation of SCM in construction, depending on whether the focus is on the supply to the site, the construction site, or both. Levels of implementation are not mutually exclusive, and in fact, they are often pursued jointly. The alternatives are numbered in Figure 2.3

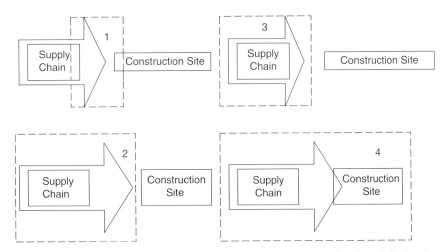

Figure 2.3 The four roles of SCM in construction. Source: Vrijhoef and Koskela (2000). Copyright Elsevier 2000.

Part A

and the descriptions below paraphrase them. One or several supply chain (SC) participants could lead each level of implementation.

- SCM focuses on the impact of the SC on construction site activities and aims to reduce the cost and duration of those activities. The primary concern, therefore, is to establish a reliable flow of materials and labour to the site.
- SCM focuses on the SC itself and aims to reduce costs, especially those related to logistics, lead time, and inventory.
- SCM focuses on transferring activities from the site to earlier stages in the SC.
- SCM focuses on the integrated management and improvement of the SC and site production, that is, site production is subsumed by SCM.

The application of SCM to the construction industry requires a huge effort. It entails developing vertical integration in the design and production process and operations to link the process into a chain focusing on maximising opportunities to add value while minimising total cost. As this application requires a significant shift in the mind-set of the participants towards collaboration, teamwork and mutual benefits, it is hardly surprising that only few sophisticated applications have been reported in the construction industry (Saad *et al.* 2001).

The extent to which any supply chain performs its functions depends very much on its nature, and the market in which it operates. Hoekstra and Romme (1992, p. 7) provided the following classification of supply chains:

- *Make-and-ship-to-stock*. Products are manufactured and distributed to stock points, which are spread out and located close to the customer.
- *Make-to-stock*. End products are made and held in stock at the end of the production process and from there are sent directly to customers who are scattered geographically.
- *Assemble-to-order*. Only system elements or subsystems are held in stock in the manufacturing centre, and the final assembly takes place on the basis of a specific customer order.
- *Make-to-order*. Only raw materials and components are kept in stock, and each order for a customer is a specific project.
- *Purchase-and-make-to-order*. No stocks are kept at all, and purchasing takes place on the basis of the specific customer order; furthermore, the whole project is carried out for one specific customer.

These different classes correlate quite well with different types of market, ranging from the large volume markets such as retail to 'one-of-a-kind' markets such as the construction industry. According to Anumba *et al.* (2000), in the large volume markets it is the supply issues rather than the product development issues which assume greater importance most of the time, whereas in the one-of-a-kind markets the focus is on the product development issues. As a consequence to this, the structure of the company will necessarily have to adapt to these priorities. For example, the large supermarket chains in the UK, in general, have a much larger proportion of resources devoted to supply and forecasting, as opposed to product development, whereas engineering companies in the construction industry have a much larger proportion of resources devoted to engineering and project management.

There is a tendency for some construction professionals, and in particular contractors, to focus on capital (or initial construction) cost and it is this focus which underpins and dominates the strategies adopted in managing the existing supply chain. Whilst capital cost is not irrelevant, most clients usually focus upon the value of their project in terms of the business case. It is this value which will form the key success factor for the project. The value may relate to the performance of the new facility in terms of its function or its worth as an asset. Additionally, it may have a value in the market place at a particular time or over many years. There is a tendency for the long-term objectives of project worth or value to become refocused on the short-term objectives of cost and time once the project or strategic brief has been established (for a discussion on this see Ashworth and Hogg 2000).

In SCM, real competition is not between organisations but rather between supply chains. In this way organisations seek to make the supply chain as a whole more competitive through the value it adds and the costs it reduces overall. Reducing cost and improving value are both important aspects of gaining competitive advantage. The traditional procurement approach sought to achieve cost reductions or profit improvement at the expense of other parties in the supply chain. The approach of transferring costs upstream or downstream does not add any competitive advantage as all costs are ultimately passed to the end user (Jones and Saad 2003, p. 231).

2.7   Supply Chain Challenges

According to Jones and Saad (2003, p. 260) there are substantial difficulties in applying SCM in the construction industry. Factors such as short-termism, lack of trust and adversarial relationships, the transient nature of construction projects and the considerable number of infrequent clients were highlighted as the main problems associated with the implementation of SCM in construction. This section will address the main challenges associated with adopting SCM in the construction industry.

There is growing interest among major clients and contractors in the UK construction industry in developing collaborative relationships (Smyth and Pryke, 2008). So far these efforts have not been very successful, although, and according to Cox and Thompson (1997, p. 129) 'the search for more collaborative relationships has become a contemporary theme in the industry'.

As the main objective of SCM is to enhance mutual competitive advantage through improved relationships, integrated processes and increased customer focus, SCM may well help improve the construction industry with its poor relationships, fragmented processes and lack of internal and external customer focus. However, there remain a number of critical issues within the construction industry that need to be considered and rectified. A long list of problems could be itemised, including lack of trust and commitment, co-ordination problems and training problems, all of which are already well documented by Latham 1994 and Egan 1998 reports. Consequently, scope for implementing SCM within construction could be limited (Jones and Saad 2003, p. 236).

One key aspect of the management of supply is the involvement of construction specialists in the design process. Most of the work involved in a construction project is carried out by specialists, usually employed as subcontractors to a main contractor. It is at this design stage that the involvement of specialists can be most beneficial - when design is carried out and cost and time parameters set. Because of the contractual link between these specialists and the main contractor (and the consequently relatively late appointment of subcontractors), subcontractors are sometimes precluded from offering advice to the client's team at an early stage. Although it is possible, through a process of nomination, to have some specialist involvement, there is a limit to the extent that nomination can realistically be achieved without exposing a client to increased risk. In a traditional (design-bid-build) construction project the only management of supply that is carried out, is carried out by the main contractor through a subcontracting process. This is usually even more price-focused than the initial contractor's tender, as the contractor seeks to negotiate an improved financial position for himself with the subcontractors he intends to employ. Consequently, the dominant traditional approach to construction procurement does not involve anything other than management of supply by contract specification, for a competitive price.

If the benefits of specialist involvement in the design process, increased innovation and collaboration, are to be achieved through supply management, traditional processes and attitudes will have to be abandoned with a new focus on the value of the project to the client.

2.8   Conclusion

It can be argued that the construction industry is not a single industry, as it rarely displays the characteristics of other industries. For example, there is no public face or single lobby. Whilst vital to economic growth, it rarely collaborates to improve either product or productivity, consisting as it does of segregated suppliers focused predominantly upon their own interests, which are often different from those of the purchasing client. Also, purchasing clients are frequently occasional and inexperienced purchasers of construction, whilst design is segregated from production by contractual terms in most cases.

To use the term 'SCM' in the context of the current UK construction industry suggests that it is possible to adopt those practices which have proved to be successful in the manufacturing and retail sectors without adapting them to reflect the particular nature of the industry and its culture. The context in the construction industry is, however, quite different. In most cases, transactions are neither ongoing nor frequent, and projects are usually unique and one-off in character. There is no production line. Many projects are procured by inexperienced purchasers and constructed by numerous specialists who have little or no contact with that purchaser. The business cultures of manufacturing and construction are also quite different.

Consequently to use the term 'supply chain management in construction' is at the very least curious, and probably simply inaccurate if it is used in the manufacturing context. Perhaps *supply networks* in construction is a more accurate phrase, suggesting a less permanent or secure relationship. SCM is conceptually ambiguous and there are problems in transferring this manufacturing-orientated management approach across to a construction industry which has characteristics that do not ideally lend themselves to SCM. This should not be a barrier to exploring the strengths of some aspects of SCM available to construction, an industry or multi-industry sector trying desperately to recast itself in a new non-adversarial, problems-solving, innovative and collaborative role.

References

Anumba, C.J., Bouchlaghem, N.M. and Whyte, J. (2000) Perspectives on an integrated construction project model. *International Journal of Co-operative Information Systems*, 9(3), 283–313.

Ashworth, A. and Hogg, K. (2000) *Added Value in Design and Construction*, London, Longman.

(BERR) Department for Business Enterprise and Regulatory Reform. (2007) *Construction Statistics Annual 2007*. London: TSO.

Briscoe, G., Dainty, A. and Millett, S. (2001) Construction supply chain partnerships: skills, knowledge and attitudinal Requirements. *European Journal of Purchasing and Supply Management*, 7(4), December, 243–255.

Cartlidge, D. (2004) *Procurement of Built Assets*. Oxford, Elsevier Butterworth-Heinemann.

Christopher, M. (1992) *Logistics and Supply Chain Management – Strategies for Reducing Cost and Improving Services*. (2nd ed.). London, Financial Times Professional Ltd.

Christopher, M. (2005) *Logistics and Supply Chain Management: Creating Value-Adding Networks*. (3rd ed.). New York, Financial Times Prentice Hall.

Cooke, B. and Williams, P. (2004) *Construction Planning, Programming and Control*. (2nd ed.). Oxford, Blackwell Publishing.

Cooper, M.C., Lambert, D.M. and Pagh, J.D. (1997) Supply chain management: more than a new name for logistics. *The International Journal of Logistics Management*, 8(1), 1–13.

The Council of Supply Chain Management Professionals (CSCMP). Available from http://www.cscmp.org/Website/AboutCSCMP/Definitions/Definitions.asp. [Accessed 8 May, 2007.]

Cox, A. and Thompson, I. (1997) Fit for purpose contractual relations: determining a theoretical framework for construction projects. *European Journal of Purchasing and Supply Management*, 3(3), 127–135.

Cox, A. and Townsend, M. (1998.) *Strategic Procurement in Construction: Towards Better Practice in the Management of Construction Supply Chains*. London, Thomas Telford Publishing.

Croom, S., Romano, P. and Giannakis, M. (2000) Supply chain management: an nalytical framework for critical literature review. *European Journal of Purchasing and Supply Management*, 6(1) March, 67–83.

Egan, J. Sir (1998) *Rethinking Construction: The Report of the Construction Task Force to the Deputy Prime Minister*. Department of the Environment, Transport and the Regions, Norwich.

Geir, G., Marianne, J. and Gøran, P. (2006) Supply chain management – back to the future? *International Journal of Physical Distribution and Logistics Management* 36(8), 643–659.

Gray, C. (1996) *Value for Money*. Reading Construction Forum and The Reading Production Engineering Group, Berkshire.

Griffith, A., Knight, A. and King, A. (2003) *Improved Tender Evaluation in Design and Build Projects*. London, Thomas Telford Publishing.

Harland, C.M. (1996) Supply chain management: relationships, chains and networks. *British Journal of Management*. 7(Special Issue), pp. 63–80.

Harland, C.M., Lamming, R.C. and Cousins, P.D. (1999) Developing the concept of supply strategy. *International Journal of Operations and Production Management*, 19(7), 650–673.

Hoekstra, S. and Romme, J. (1992) *Internal Logistic Structures: Developing Customer-Oriented Goods Flow*. London, McGraw-Hill.

Jones, M. and Saad, M. (2003) *Managing Innovation in Construction*. London, Thomas Telford Publishing.

Latham, M. Sir (1994) *Constructing the Team: Final Report of the Government/Industry* Review of Procurement and Contractual Arrangements in the UK Construction *Industry*. London, HMSO.

Long, D. (2004) *International Logistics: Global Supply Chain Management*. Dordrecht, Kluwer Academic Publishers.

Lummus, R.R., Krumwiede, D.W. and Vokurka, R.J. (2001) The relationship of logistics to supply chain management: developing a common industry definition. *Industrial Management and Data Systems*, 101(8), 426–432.

McCormack, K., Ladeira, M. and de Oliveira, M. (2004) The development of a SCM process maturity model using the concepts of business process orientation. *Supply Chain Management: An International Journal*. 9(4), 272–278. Pearson, A. (1999) Chain reaction. *Building*. 10, 54–55.

Porter, M.E. (1985) *Competitive Advantage: Creating and Sustaining Superior Performance*. New York, NY Free Press.

Royal Institution of Chartered Surveyors. (2006) *Contracts in Use: A Survey of Building Contracts in Use During 2004*. London: Davis Langdon.

Saad, M., Jones, M. and James, P. (2001) A review of the progress towards the adoption of supply chain management (SCM) relationships in construction. *European Journal of Purchasing and Supply Management*, 8(3), September, 173–183.

Shingo, S. (1988) *Non-Stock Production Non-Stock Production: The Shingo System for Continuous Improvement*. Cambridge, Productivity Press.

Smyth, H. and Pryke, S.D. (2008) *Collaborative Relationships in Construction: Issues and Experiences*. Oxford, Blackwell.

Supply Chain Council. Available from http://www.supply-chain.org [Accessed 28 April, 2007].

UK SIC. (2003) *UK Standard Industrial Classification (SIC) of Economic Activities 2003*. London: The Stationery Office. Available from <statistics.gov.uk/methods_quality/sic/downloads/UK_SIC_Vol1(2003).pdf> [Accessed 28 April, 2007].

Van der Veen, J. and Robben, H. (1997) Supply chain management: een overzicht. supply chain management: an overview. *Nijenrode Mgmt. Review*, 6, 62–75.

Vrijhoef, R. and Koskela, L. (2000) The four roles of supply chain management in construction. Supply Chain Management in construction – special issue. *European Journal of Purchasing and Supply Management*, 6(3), December, 169–178.

Part A

3

Culture in Supply Chains

Richard Fellows

'Essentially, business is about appropriating value for oneself . . . only by having the ability to appropriate value from relationships with others . . . can business be sustained . . . [there] . . . must . . . be conflicts of interest between vertical participants in supply chains, just as there are between those competing horizontally . . . In Western . . . culture most suppliers are basically opportunistic'. (Cox, 1999; p. 171). Most clients, consultants and constructors are businesses and so, operate under business performance imperatives. The businesses are subject to regulation and, particularly for (design) consultants, the need to behave professionally – on a moral/ethical basis for perceived social good.

While the concepts of competition are applied to horizontal situations (competitors in similar positions within other supply chains), most commonly, vertical competitions between members of a supply chain (those at different positions, or tiers, within the supply chain) may have much greater consequences for the effectiveness of the output and the efficiency of the supply processes. A systems model of the typical realisation process of a construction project is shown as Figure 3.1. The larger and the more complex the project, the greater is the number and diversity of specialist participants in each of the functional categories. Internationalisation and globalisation compound the ensuing difficulties.

Usually, there is a great diversity of participants and specialist functions. Therefore, it is difficult, if not impossible, to identify all the participants at the early stages of a project , prompting assumptions of identities and interests of parties not involved at this point; it can be very tempting, for simplicity and convenience, to ignore such parties!

Strict application of the concept of supply chains may give a somewhat limited perspective of the realities of project realisation, occupation and use, and disposal processes. Focusing on project realisation, it seems more appropriate to examine networks of potential participants, leading to integrations of chains of suppliers for actual realisation processes, preferably

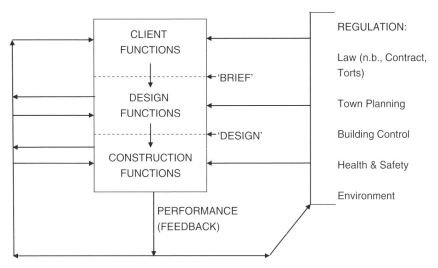

Part A

Figure 3.1 The project realisation process. *Note*: (1) Performance leads to satisfaction of participants and, hence, (perspectives of) project success. (2) Performance-Satisfaction-Success also produces feedforward in the 'cycling' of project data and information to aid realisations of future projects through participants' perception-memory-recall filtering ('experiences').

incorporating 'client' purpose-oriented perspectives also. This encourages use of the concept of 'value-added' by the supply chain (network) members. Some chains operate in series; others, in parallel.

Supply networks operate in two primary contexts. First, as horizontal arrays of organisations from which particular organisations may be selected to participate in the realisation of a project, depending on factors such as expertise, resources, workload and price. Second, the selected participants then form a vertically oriented network of organisations with their integration aimed at realising the project effectively and efficiently. In these contexts, much of the research and literature has focused on the realisation processes with the objective of enhancing *project management performance* (time, cost and quality measures) (see, e.g. Jin, Doloi and Gao, 2007); the design, briefing, and buildability/constructability literature adopts a focus more towards *project performance* (functional effectiveness of the project in occupation and use) (see, e.g., Green and Simister, 1999; Griffith and Sidwell, 1995).

Thus, the construction project business world emphasises operating/ performance imperatives relating to individual organisations – the links in the construction supply chain/network. However, the holistic performance perspectives (which are becoming ever more widespread) to yield successful projects with satisfied participants emphasise integration of the supply network members.

Thorelli (1986) suggests that networks lie between markets and hierarchies, as quasi-firms or multi-organisations. That is part of a contingent perspective on firms' macro-behaviour, in that they operate in complex environments in which firms' behaviours are interdependent; the degree of interdependence being contingent on the environment. Thus, competing becomes an issue of a firm's locating itself within the network, as in a Cournot equilibrium under oligopoly.

Cox (1999) distinguishes supply chains, which deliver physical goods and services to customers, from value chains which relate to the revenue streams from customers and so, operate in the reverse direction. Such a perspective limits conceptualisation and analysis due to its market price orientation. Whilst appreciation of the flows of goods/services and the revenues generated from them is useful, understanding the fundamental causations of the flows, and amounts of those flows, is vital for systemic improvements.

The usual objective of business transactions is increasing the wealth of the participants. The chaining and networking concepts provide the linkages between the individual participants – why and how they relate to each other for realisation of their objectives. Thus, participants need to '. . . understand the physical resources that are required within a supply chain to create and deliver a finished product or service to a customer. . . . understand the exchange relationship between particular supply chain resources and the flow of revenue in the value chain. . . . understand what it is about the ownership and control of particular supply chain resources to command more of the flow of value than others' (Cox, 1999). Such appropriation of value depends, to a major extent, on the power and relational structures of the supply chain.

Usually, wealth is measured in terms of financially realisable assets. That objective is the underpinning of market capitalism, the generic economic system of most societies. However, the world retains a rich mix of 'not-for-profit organisations' which, although requiring financial break-even performance, measure wealth in various ways and, often, for several stakeholder groups. Such a situation is also occurring amongst for-profit organisations, which embrace various non-financial performance measures to supplement the financial ones.

What such contextual analyses indicate is that the organisations are operating in a world of increasing pluralism and, hence, complexity which is driven by, and, in turn drives, culture and cultural change. Wealth is determined by what people value and so the pluralistic approach to wealth manifests the diversity of people's values. Those values are articulated through the manifestations of culture by governing how people behave and conduct relationships and with what consequences for performance of individual and collective activities (such as project realisation).

Thus, this chapter addresses the nature of societal culture, both national and organisational, together with derivatives of organisational climate and the behavioural concerns of behaviour modification, organisational citizenship and corporate social responsibility. Consideration of ethics addresses the moral bases which impact on human behaviour. Relational issues of team formation and functioning and alliances between businesses are examined as human and organisational contexts through which supply chain participants operate and projects are realised.

Typically, culturally oriented analyses have been used to determine the appropriate dimensions for study and to detect typical behaviours and their

Part A

underpinning causes (values and beliefs – see below). Those studies have facilitated further work to examine interrelationships between human groups (nations, organisations, etc.), commonly featuring the examination of cultural differences and/or measures of cultural distance. Such examinations have been applied to strategic alliances between organisations and so, are of direct importance to the inevitable joint-venture nature of construction supply chains (whether for domestic or international projects). Inevitably, analyses have concentrated on formal organisations and alliances whilst the more common encounters, by far, within construction are informal.

Thus, in construction supply chains/networks, two organisations impact on each participant directly – the organisation which employs that person and the project multi-organisation to which that person is attached to fulfil a supply chain role. Of course, the diversity of construction project members and stakeholders (both individuals and organisations) with which any particular member of the project supply chain must interact lends further complexity to appreciation of behavioural imperatives of the supply processes to secure successful performance which requires understanding of and sensitivity to accommodate diverse, often competing, interests.

The argument advanced here is that only through appreciation of the requirements of others can fragmented supply chains (individual objectives) be appropriately integrated to secure both individual and holistic successful performance of both the project management process (realisation) and the project (product in use). That appreciation relates to own and others' cultures in terms both of the manifestations of the cultures (language, behaviour, etc.) and the underpinning/determining variables and constructs (values, beliefs). That appreciation should include organisational climates, what drives behaviour and how behaviour may be changed together with the likely outcomes of any change initiatives.

Such appreciation is essential for management of construction supply networks/chains due to the wide array of diverse stakeholders and the labour intensive nature of the processes. Intra-organisational management, inter- organisational management and boundary management are intensified in the fluid, power-based temporary multi-organisations which are the norm for any construction project.

3.2   Culture

Initially, culture may be described as 'how we do things around here' (Schneider, 2000). Of course, much more is involved: culture is not merely *how* things are done, the scope is much more extensive and includes *what* is done, *why* things are done, *when*, and *by whom*. . . . However, the description does have a behavioural focus and so draws attention to that primary manifestation of culture.

Kroeber and Kluckhohn (1952) define culture as, '. . . patterns, explicit and implicit of and for human behaviour acquired and transmitted by symbols, constituting the distinctive achievements of human groups,

Part A

including their embodiment in artefacts; the essential core of culture consists of traditional (i.e. historically derived and selected) ideas and, especially, their attached values; culture systems may, on the one hand, be considered as products of action, on the other as conditioning elements of future action.' Culture is a collective construct which concerns groups of people. Further, culture is iteratively dynamic – culture shapes behaviour and, in turn, behaviour shapes culture; development spirals through time.

Hatch (1993) advances a model of cultural dynamics which encapsulates the cyclical processes of manifestation, realisation, symbolisation and inter-pretation. The dynamism arises from the continual construction and recon-struction of culture as the context for setting goals, taking action, making meaning, constructing images and forming identities. In construction, a vital consideration is the impact of culture on what performance is achieved and measured against pre-determined, culturally bound, targets.

Hofstede (1994a) defines culture as '. . . the collective programming of the mind which distinguishes one category of people from another.' This defini-tion suggests that culture is learned, rather than being innate in the person or genetic. It is inherited behaviourally through replicating and responding to the behaviour of others, most importantly close family and contacts during early life. Further, as culture is a collective construct, categorisation of people may be by ethnic origin, political nationality, organisation, etc. (the important aspect of such categorisation is that 'within category' vari-ability is significantly less than 'between category' variability; although, for certain categorisations in practice – notably, nation-states – within category variation may be large). This is very important in examining similarities and differences between cultures – where the boundaries are drawn, what dimensions are considered, and how measurements are made and used (absolute/relative).

Schein (1990) views culture as grounded in basic assumptions which constitute communal values and are 'taken for granted'. Cultures arise through the formation of norms of behaviour relating to critical incidents which are communicated through stories passed on between members of the community, as well as through identification with leaders and what they scrutinise, measure and control.

Many models of culture employ vertical analyses: physiological instincts and beliefs at the core (survival imperatives; religion, morality, etc.); values as the intermediate layer (the hierarchical ordering of beliefs, perhaps with possible trade-offs); and behaviour at the outer layer (as in language, symbols, heroes, practices, artefacts), as in Figure 3.2.

Culture may be analysed horizontally also, which yields categories of national culture, organisation culture, organisation climate, and behaviour of people (see Figure 3.3). Due to interrelationships (caused by people fulfill-ing different roles), the categorical boundaries are quite fuzzy. Individuals, at any one time, may be of a certain nationality (e.g. Chinese), work in a particular organisation (e.g. Bechtel), belong to a 'social interest group' (e.g. Greenpeace), and show certain behaviour (e.g. organisational citizen-ship behaviour). Clearly, situational variables impact on behaviour, promot-ing a contingency perspective. The complexity has prompted a variety of

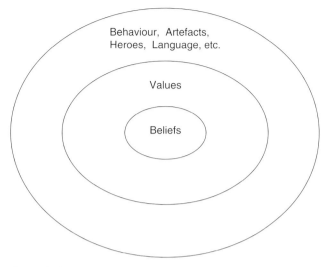

Figure 3.2 'Layers' of culture. *Note*: Schein (2004) considers levels of culture to be artefacts, espoused beliefs and values, and underlying assumptions (as the deepest level).

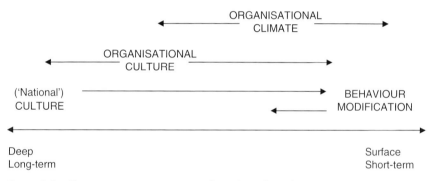

Figure 3.3 Change spectrum. Source: Fellows (2006). With permission from the Chartered Institute of Building. *Note*: Boundaries between cultures, climates and behavioural modifications are fuzzy.

approaches for determining and analysing the categories. One consequence is a common confusion of the categories when 'managers' wish to effect change – in particular what has changed, how, why, and the permanence of any change – culture cannot be changed by use of a '40-hour workshop' (although behaviour modifications may result).

3.3   Dimensions of Culture

In order to gain appreciation of culture and to understand similarities and differences between cultures, it is helpful to determine dimensions on which cultures may be 'measured' – most advisedly, employing relative positionings

(Hofstede, 2001). The dimensions of culture and related constructs constitute common bases for determining 'profiles' of cultures, to facilitate both appreciation of individual cultures, as well as comparisons of cultures (which, commonly focus on differences).

In order to assemble a comprehensive perspective of culture, the levels of national culture, organisational culture, organisational climate and behaviour modification are examined (i.e., progression from the more general to the particular). The principles and knowledge of cultures are examined to enable a supply chain manager to gain understanding of the manifestations likely to be encountered in project practice and to determine actions from an informed basis. Clearly, the number of possible combinations existing in the world is enormous!

3.3.1   National culture

Hofstede (1980) determines four dimensions for measuring national culture:

- Power Distance – 'the extent to which the less powerful members of institutions and organizations within a country expect and accept that power is distributed unequally'. (Hofstede, 1994b: 28)
- Individualism/Collectivism – 'Individualism pertains to societies in which the ties between individuals are loose: everyone is expected to look after himself or herself and his or her immediate family. Collectivism as its opposite pertains to societies in which people from birth onwards are integrated into strong, cohesive in groups, which throughout people's lifetimes continue to protect them in exchange for unquestioning loyalty.' (ibid: 51)
- Masculinity/Femininity – 'masculinity pertains to societies in which gender roles are clearly distinct (i.e., men are supposed to be assertive, tough, and focused on material success whereas women are supposed to be more modest, tender, and concerned with the quality of life); femininity pertains to those societies in which social gender roles overlap (i.e., both men and women are supposed to be modest, tender, and concerned with the quality of life).' (ibid: 82–83)
- Uncertainty Avoidance – 'the extent to which the members of a culture feel threatened by uncertain or unknown situations.' (ibid: 113)

A fifth dimension of Long-Termism/Short-Termism was added later (Hofstede, 1994b), following studies in Asia which detected important impacts of 'Confucian Dynamism' (The Chinese Culture Connection, 1987). Long-Termism is 'the fostering of virtues orientated towards future rewards, in particular perseverance and thrift.' (ibid: 261), whilst Short-Termism is 'the fostering of virtues related to the past and present, in particular respect for tradition, preservation of 'face', and fulfilling social obligations.' (ibid: 262–3). Other researchers, e.g. Trompenaars and Hampden-Turner (1997; 123–5), recognise the culturally bound approaches to management of time, characterised as sequential or synchronic behaviour. Sequential behaviour leads people to deal with matters one at a time, often involving prioritising them by perceived importance; punctuality is vital (common in 'Western' cultures). Synchronic behaviour occurs where people deal with matters as they arise, often involving 'multi-tasking' (for instance, immediate greeting of a visitor is essential, even if another vital task is being carried out when the visitor arrives); punctuality is not too important (common in 'Eastern' cultures).

Gomez, Kirkman and Shapiro (2000) explain that people in collectivist cultures favour in-group members but discriminate against out-group members. Chen, Meindl and Hunt (1997) examine the cultural dimension of collectivism and determine that it is a construct which comprises vertical and horizontal components. They juxtapose those components to Hofstede's (1980) concept of individualism as, '. . . individualism (low concern for collectivity and low concern for in-group others) at one end of the spectrum with vertical collectivism (high concern for the collectivity) and horizontal collectivism (high concern for in-group others) at the other end'. They find that, 'Because the vertical scale items refer to work situations and the horizontal scale items primarily refer to non-work situations, one may speculate that the Chinese are becoming 'organizational individualists' even though they are still cultural collectivists in other domains . . .'. This finding may be extended to the 'Asian Tiger economies', consequential upon their rapidly rising levels of industrialisation and wealth (e.g. Triandis, 1990; Hofstede, 1983, 1994b:75). Hofstede (1983) notes the correlation between wealth and Individualism in various countries but continues that '. . . collectivist countries always show large Power Distances but Individualist countries do not always show small Power Distance'. These relationships are discussed further in Hofstede (2001).

3.3.2   Organisational culture

Usually, organisational cultures derive and evolve from the founders and others who have had major impact on the organisation's development. Such people, through setting organisational objectives and influence over employment of staff, have shaped the values and behaviour of members of the organisation to develop the organisation's identity – both internally and externally. Organisational cultures (and climates) are, largely, self-perpetuating – persons who 'fit' are hired and they 'fit' because they are hired; errors of 'fit' are subject to resignations or dismissals. Thus, organisational cultures develop through the necessity of maintaining effective and efficient working relationships amongst stakeholders (both permanent and temporary). Pressure for cultural change commonly arises from external parties, particularly in situations of environmental turbulence, innovations, and attempts to enter new markets.

Organisational culture types and the dimensions determined for analyses (see below) show marked categorical similarity to the organisational behaviour typology of the human relations – task schools (such as Herzberg, Mausner and Bloch Snyderman, 1967 – theory X and theory Y). This dichotomous perspective emphasises the spectrum of, often alternative, orientation of an organisation towards people ('employees', as the 'active factor' in supply chain processes which, as such, are the primary determinants of performance and, hence, success) or outcome (the product/service output as the 'supply' of the organisation). Today, the concept of 'employees', especially in a supply chain perspective, should be interpreted liberally to encompass all types and levels of persons employed in an organisation, whether a permanent organisation (e.g. a company) or a temporary (multi-) organisation (as for a construction project supply chain) and with a view to including perspectives on and behaviour towards other stakeholders – as conceived in corporate social responsibility (CSR) and organisational citizenship behaviour (OCB); the outward and inward manifestations of organisational ethics.

Schein (1984) determines two primary types of organisational culture: 'free flowing', an unbounded, egalitarian organisation without much formal structure, thereby encouraging debate and internal competition; 'structured', a bounded, rigid organisation with clear rules and requirements (analogous to the organicn–mechanistic typology of Burns and Stalker, 1961).

Handy (1985) suggests that the main factors which influence organisational culture are: history and ownership, size, technology, goals and objectives, environment and people. He advances four primary forms of organisational culture (echoed by Williams, Dobson and Walters, 1989).

- *Power* is a web with the primary power at the centre; emphasis is on control over both subordinates and external factors (suppliers etc. and nature).
- *Role* involves functions/professions which provide support of the overarching top management; emphasis is on rules, hierarchy and status through legality, legitimacy and responsibility (as in contractual rights, duties and recourse).
- *Task* in which jobs or projects are a primary focus, yields an organisational net (as in a matrix organisation); structures, functions, and activities are evaluated in terms of contribution to the organisation's objectives.
- *Person* in which people interact and cluster relatively freely; emphasis is on serving the needs of members of the organisation through consensus.
  - Hofstede (1994b, 2001) identifies six dimensions of organisational cultures: Process–Results Orientation (technical and bureaucratic routines (can be diverse); outcomes {tend to be homogeneous}). This reflects means or ends orientation. Process cultures tend to be routine based and (developed to be) risk-avoiding. Strong (homogeneous) cultures tend to be more results oriented.
  - Job–Employee Orientation (derives from societal culture as well as influences of founders, managers). Job cultures emphasise getting a job done (only outputs of employees matters), whilst employee cultures focus on concern for the welfare of the people involved (including personal matters).

Part A

○ Professional–Parochial (one category of people identify with a profession(s), their type of job/occupation; other people identify with employing organisation). People in parochial cultures consider that the norms of the organisation apply outside the workplace, hiring is on an holistic perspective and the firm takes the long-term view, allowing individuals to be more short term in focus. Professional people separate private life and work life aspects, they are hired for occupational competence and adopt long-term perspectives.

○ Open–Closed System (ease of admitting new people, styles and ease of internal and external communications). In open cultures, new people are fully and rapidly incorporated, whilst in closed systems inclusion is likely to take a very long time and such systems remain highly secretive towards people inside the organisation as well as towards those outside it.

○ Tight–Loose Control (degrees of formality, punctuality, etc., may depend on technology and rate of change). Tight control demands extensive and rigid structuring with high levels of cost consciousness and time-keeping; stringent, if unwritten, codes of behaviour and dress follow.

○ Pragmatic–Normative (how to relate to the environment, n.b. customers; pragmatism encourages flexibility). Pragmatic organisations are driven by markets, usually emphasising customer orientation. Normative organisations emphasise following rules and procedures and are perceived as having high standards of honesty and ethics. Depending on the nature of the business and its context, either approach may lead to good performance (although persons in pragmatic organisations see themselves as more results oriented).

Cameron and Quinn (1999) employ a 'competing values' model which juxtaposes 'flexibility and discretion' and 'stability and control' on one dimension, with 'internal focus and integration' and 'external focus and differentiation' on the other. The resultant model (see Figure 3.4) yields four quadrants, each denoting a type of organisational culture – Clan, Adhocracy, Market, and Hierarchy.

• Clan – 'Some basic assumptions in a clan culture are that the environment can be best managed through teamwork and employee development, customers are best thought of as partners, the organization is in the business of developing a humane work environment, and the major task of management is to empower employees and facilitate their participation, commitment, and loyalty.' (ibid: 37)

• Adhocracy – 'A major goal of an adhocracy is to foster adaptability, flexibility and creativity where uncertainty, ambiguity and/or information-overload are typical. Effective leadership is visionary, innovative and risk-orientated. The emphasis is on being at the leading edge of new knowledge, products, and/or services. Readiness for change and meeting new challenges are important.' (ibid: 38–9)

• Market – 'The major focus of markets is to conduct transactions with other constituencies to create competitive advantage. Profitability, bottom

```
                    ┌───────────────┐
                    │  FLEXIBILITY  │
                    │      AND      │
                    │  DISCRETION   │
                    └───────────────┘
```

CLAN CULTURE

Leader:
Facilitator; Mentor; Parent

Effectiveness Criteria:
Cohesion; Morale;
Development of Human
Resources

Organisation Theory Basis:
Participation fosters
commitment

ADHOCRACY CULTURE

Leader:
Innovator; Entrepreneur;
Visionary

Effectiveness Criteria:
Cutting-edge output;
Creativity; Growth

Organisation Theory Basis:
Innovation fosters
new resources

```
┌───────────────┐                    ┌───────────────┐
│INTERNAL FOCUS │                    │EXTERNAL FOCUS │
│      AND      │                    │      AND      │
│ INTEGRATION   │                    │DIFFERENTIATION│
└───────────────┘                    └───────────────┘
```

HIERARCHY CULTURE

Leader:
Coordinator, Monitor,
Organiser

Effectiveness Criteria:
Efficiency; Timeliness;
Smooth functioning

Organisation Theory Basis:
Control fosters
efficiency

MARKET CULTURE

Leader:
Hard-driver; Competitor;
Producer

Effectiveness Criteria:
Market share; Goal
achievement; Beating
competitors

Organisation Theory Basis:
Competition fosters
productivity

```
                    ┌───────────────┐
                    │   STABILITY   │
                    │      AND      │
                    │    CONTROL    │
                    └───────────────┘
```

Figure 3.4 Competing values and organisational cultures model (following Cameron and Quinn, 1999).

line results, strength in market niches, stretch targets, and secure customer bases are primary objectives for the organization. Not surprisingly, the core values that dominate market type organizations are competitiveness and productivity.' (ibid: 35)
- Hierarchy – 'The organizational culture compatible with this form is characterised by a formalized and structured place to work. Procedures govern what people do. Effective leaders are good coordinators and organizers. Maintaining a smooth-running organization is important.

The long-term concerns of the organization are stability, predictability, and efficiency. Formal rules and policy hold the organization together.' (ibid: 34)

3.3.3   Organisational climate

A common difficulty is to differentiate organisational culture and organisational climate. Mullins (2002: 906) explains organisational climate as: 'Relating to the prevailing atmosphere surrounding the organisation, to the level of morale, and to the strength of feelings or belonging, care and goodwill among members. Organisational climate is based on the perceptions of members towards the organisation.' Organisational climate operates between organisational culture and organisational behaviour and so, may change more quickly than organisational culture, but far less rapidly than organisational behaviour.

Victor and Cullen (1988) discuss organisational climate at two levels. The first level is aggregate perceptions of organisational conventions concerning forms of structure and procedures for rewards and control (perceptions of practices and procedures – Schneider, 1975). The second level is aggregate perceptions of organisational norms concerning warmth towards and support for peers and subordinates (organisational values – Denison, 1996; Ashforth, 1985).

'Organizational Climate is a relatively enduring quality of the internal environment of an organization that (a) is experienced by its members, (b) influences their behaviour, and (c) can be described in terms of the values of a particular set of characteristics (or attributes) of the organization.' (Tagiuri and Litwin, 1968: 27). Thus, the climate of an organisation distinguishes it from other, similar organisations. As shared experience of members of an organisation, it reflects their perceptions about autonomy, trust, cohesion, fairness, recognition, support, and innovation and so leads to the members of the organisation having shared knowledge and meanings. Organisations' climates are important contributors to homogeneity amongst members.

Organisational climate, and its groundings in organisational culture and national culture, is important in establishing the organisation's identity, as strongly manifested in its (business) objectives and behaviour – both of the organisational entity and of its individual agents. Given the strongly prevailing market competitive business paradigm, which tends to reinforce and perpetuate self-oriented (individualistic) behaviour, which seems to be increasingly short-term also (Hutton, 1996, 2002), there are antithetical pressures on organisations and their agents in supply chains to produce profits in support of dividend imperatives, whilst behaving as partners with others in the supply chain and operating under long-term concession procurement systems with their commensurate greater costs and risks.

The delicate art for managers in such supply chains is to secure appropriate balancing of interests and rewards, for which it is essential to carry the other actors along a path for common anticipated benefit. The mechanisms

Part A

for such 'balancing acts' are founded in understanding and being sensitive to incorporating others' requirements with their own requirements. That requires thorough appreciation of the organisational context and how, in relation to the project, it impacts on the behaviour of the actors through influence on their motivation and commitment.

3.3.4   Behaviour modification (BMod)

Modifying the behaviour of employees is a common endeavour of motivational schemes – usually associated with positive, systematic reinforcement, via rewards, of behaviour which enhances productivity. Provision of the reward is contingent on both behavioural change and, consequently, enhanced performance. It is the link between the BMod and the increase in output of the employees which is important and, so, which occasions the reinforcement reward. The reinforcements are likely to vary between people and can operate in the negative direction too – as 'punishments' for unwanted BMod.

The simplicity of the required associations in BMod have led to its acceptance being limited, as well as to issues of ethics relating to employee choice. However, a significant aspect of BMod is that the causal chain operates rapidly. A further, potential, detrimental effect is that the effectiveness (e.g., financial rewards as reinforcement) may be only temporary. In using legislation to effect BMod, sustaining the legislation, coupled with a programme of education, can lead to culture change by amending people's beliefs about their practices – important for industrial safety, for example.

3.4   Values and Value

Usually, value is manifested in monetary terms. Such modern interpretation of exchange value is essential in market business contexts, but is also increasingly recognised as restrictive and inadequate to provide comprehensive measurement of performance. Stakeholder perspectives, which include recognition that an individual may fulfil diverse roles (organisational employee, consumer, environmentally aware person, etc.), corporate social responsibility (CSR) metrics (including environmental impacts, ethics) are becoming widespread to lend credibility to the range of values (as things of worth to people) which are important – this extends the perspective of use value.

Thus, generic value, what something is perceived to be worth (to society or to an individual decision maker) determines what is produced and how production occurs. For construction projects to be successful, it is essential to determine who the stakeholders are and what their objectives are for the project – i.e., what are their values and the structuring of those values – in order to examine compatibility and, within the parameters of the project, to secure acceptance of performance targets, derived from the values, to apply to the project. This extends the remit for briefing even beyond the strategic and project typology (e.g., Green and Simister, 1999) into the realm of determination of the values of the major stakeholders and their compatibility, as investigated by Mills, Austin and Thomson (2006).

It is the values of the project stakeholders which determine the pressures on the supply chain members of what to provide. It is quite obvious and well known that stakeholders at different vertical positions in the supply chain are likely to have different value structures. However, even though stakeholders at the same vertical level in the supply chain may have similar values, this, in itself, does not ensure collaborative, synergetic performance. It is vital not only to determine what the values are (their identities), but also how they operate in the context of the project. Indeed, different values of different stakeholders may be complementary for performance (e.g., project-functional orientation of designers and project-quality orientation of constructors), whilst similar values of different stakeholders may yield detrimental consequences (own, short-term, profit orientation). The critical issue is compatibility of the values and their practical manifestations within the project constraints.

'A transaction is the exchange of *values* between two parties. The things-of-value need not be limited to goods, services, and money; they include other resources such as time, energy and feelings' (Kotler, 1972). Rokeach (1973) considers that values are the deeply held, enduring beliefs of people; value is the benefit resulting from an exchange and arises from people's preferences. Thus, Kotler is referring to exchanges of things (tangible and/or intangible) to which the transacting parties attach values, both in exchange and in use.

Values are 'desirable, transsituational goals, varying in importance, that serve as guiding principles in people's lives' (Schwartz and Bardi, 2001). Values are positive, because they are desirable, and generic, because they are transsituational and so, are different from specific objectives, which they underpin. Schwartz and Bilsky (1987) advance 'five features that are common to most . . . definitions of values . . . (a) concepts or beliefs, (b) about desirable end states or behaviours, (c) that transcend specific situations, (d) guide selection or evaluation of behaviour and events, and (e) are ordered by relative importance.'

Values can be classified as ends (situations: outcomes – as in the functioning of a project in use) or means (instrumental values: processes – as in project realisations which consume less resource and produce less pollution).

It is usual to regard values as motivators of human behaviour (Schwartz and Bilsky, 1987 suggest nine motivational domains of values; amended to ten motivational types of values in Schwartz and Sagiv, 1995), along with needs (e.g., Maslow, 1943; Alderfer, 1972) and means (e.g., Vroom, 1964). Values refer to what people believe to be important and so, are instrumental in generating goals and targets. Schwartz and Bilsky's (1987) motivational domains of values support the perspective of congruence between people's values and those expressed for tasks/projects having a positive effect on performance.

Table 3.1 Higher order values, constituent motivational types of values, and goals. Derived from Schwartz and Bardi (2001)

Higher Order Value	Motivational Types of Values	Goals
Universalism	Broad-minded, wisdom, social justice, equality, world at peace, world of beauty, unity with nature, protecting the environment	Understanding. appreciation, tolerance, protection of the welfare of all people and of nature
Benevolence	Helpful, honest, forgiving, loyal, responsible	Preservation and enhancement of the welfare of people with whom one is in frequent personal contact
Conformity	Politeness, obedient, self-discipline, honouring parents and elders	Restraint of actions, inclinations and impulses likely to upset or harm others and violate social expectations and norms
Tradition	Humble, accept position in life, devout, respect for tradition, moderate	Respect, commitment, and acceptance of the customs and ideas that traditional culture or religion provide
Security	Family security, national security, social order, clean, reciprocation of favours	Safety, harmony, and stability of society, of relationships , and of self
Power	Social power, authority, wealth, preserving public image	Social status and prestige, control or dominance over people and resources
Achievement	Successful, capable, ambitious, influential	Personal success through demonstrating competence according to social standards
Hedonism	Pleasure, enjoying life	Pleasure and sensuous gratification for oneself
Stimulation	Daring, a varied life, an exciting life	Excitement, novelty, challenges in life
Self-Direction	Creativity, freedom, independent, curious, choosing own goals	Independent thought and action, choosing, creating, exploring

Schwartz and Sagiv (1995) and Schwartz and Bardi (2001) advance a model of motivational types of values which is derived from research into individual's values and consistency between them (Table 3.1). They discuss the organisation of the value types into two dimensions: Openness to Change versus Conservation; Self-Enhancement versus Self-Transcendence. Honesty and other 'pro-social' values are important, while power values, including wealth, are far less important. Consensus over the level of importance of hedonism values is low. Notably, the research reveals differences between value hierarchies of different occupational and national groups. This confirms the necessity for identification of the values of project stakeholders in the supply chain and, more especially, for developing frameworks to secure acceptably compatible manifestations of these values for project realisation – performance targets, etc. This is a primary task of the project (supply chain) manager to address right from the initiation of the project (project conception).

3.5   Ethics

Ethics concern human interactions – what people do, how the things are done and with what impacts on other people; as such, they are related to values very closely and constitute an important, integral component of culture. (In fact, ethics are the manifestations of moral values.) Further, ethical concerns feature ever more widely in evaluations of projects and organisations. Codes of ethics (conduct) usually indicate the boundaries of what is acceptable behaviour and being acknowledged as ethical is recognised as valuable for marketing advancement. The 'bottom line', of course, is that legal systems are grounded in morals and ethics and denote the absolute limits of acceptable behaviour – notably, there are significant differences between cultures.

A reputation for honesty and good, moral behaviour attracts business and tends to lead to reduction of transaction costs through reducing promotion and scrutiny requirements. These are important cost and organisational issues for a project supply chain manager to address.

Hinman (1997) distinguishes morals as first-order beliefs and practices about what is good and what is bad which guide behaviour; and ethics as second-order, reflective consideration of moral beliefs and practices. Rosenthal and Rosnow (1991: 231) note 'ethics *refers* to the system of moral values by which the rights and wrongs of behaviour . . . are judged' [italics added].

Issues of definition and perspective on ethics have led to the development of four primary paradigms. In deontology (relating to duty or moral obligation), a universal moral code applies. In scepticism (relativism; subjectivism), ethical rules are arbitrary and relative to culture and to time; that is extended into ethical egoism where ethics become matters for the conscience of the individual. Thus, egoism concerns pursuit of self-interest and, as such, can be related to common business performance criteria (notably, profit maximisation). Teleology (the branch of philosophy relating to 'ends' or final causes) constitutes a utilitarian approach where ethics are dependent upon the anticipated consequences – prompting a cost-benefit perspective, perhaps invoking the judgmental criterion of 'the greatest good for the greatest number' which, itself, is likely to necessitate subjectively determined weightings. Objectivism asserts that there are definitions of right and wrong which are accepted generally (either universally or more locally) (Leary, 1991: 261–262).

Thus, given the diversity of ethical paradigms, there remains great scope for variability in determination of what is ethical – a 'tip' in one context/ society may constitute a 'bribe' elsewhere. Due to the deep-seated nature of ethics and their moral foundation, the project supply chain manager should establish a code of ethical behaviour which is appropriate to the project location and to the stakeholders; that code should be documented and communicated to all members of the supply chain and adherence to it policed (with suitable sanctions for transgressions).

Ethics concerns how the actions of one person may impact on others and so, imposes a 'duty of care' not to harm others. That perspective generates questions of to whom such a duty is owed, together with concerns over whether it should be applied absolutely or relatively (deontologically or teleologically). Law and codes of conduct endeavour to denote boundaries of application (the 'neighbour' principle; the client). Clearly, national law employs wide boundaries and applies to all people in the country (the jurisdiction of the law); codes of conduct apply more restrictively and may be specific regarding behaviour towards specified others likely to be encountered in the course of activities (notably, the client of a construction – professional – consultant.)

However, stepping beyond such rather arbitrarily drawn boundaries into the realm in which a professional is a person who possesses special knowledge, which, itself, concerns generic 'good', and uses that knowledge for the benefit of the immediate client and wider society, the boundaries of application vanish. Benefit from professional activities is the objective but, at the same time, leaving distributions of such benefits open to judgement due to the, frequent diversity of people affected by a professional's work.

The prospect of 'compartmentalisation' through the presence of artificial boundaries around behavioural requirements and perspectives – as governed by circumstances (domestic, employment etc.) – prompts Fellows, Liu and Storey (2004) to discuss the notion of 'personal shielding', in which a person amends his or her (ethical) behaviour to accord with the expressed or perceived ethics of another, usually an employing, organisation. Such shielding may feature in principal-agent circumstances, including those between a commissioning client and a design consultant.

3.6   Organisational Citizenship Behaviour (OCB) and Corporate Social Responsibility (CSR)

OCB concerns the voluntary behaviour of employees towards the benefit of the organisation in excess of the requirements of both the contract of employment specifications and the norms of behaviour of similar employees (Organ, 1988); the employees of the organisation 'go the extra mile' (for the organisation's benefit). For such behaviour to occur, employees must feel committed to the organisation, which results from their own disposition and their perception of how the organisation (and it's superiors) treats them. Thus, adopting the perspective that an organisation has a personality and behaviour separate from its members (Wayne, Shore and Linden, 1997), it is appropriate to examine the reciprocations in the relationship between employees and the organisation.

For a construction project, OCB can occur within a single supply chain organisation (a firm), within the project temporary multi organisation (TMO) assembled to realise the project, or both. Although OCB is examined most often in relation to a single firm regarding relationships between the firm and persons within it, OCB's applicability may be extended, by analogy, to apply to behaviour at the inter-firm level, as for a whole supply chain/network.

Organ (1988) employs dimensions of altruism (discretionary behaviour which assists others), conscientiousness (fulfilling role requirements in excess of the minimum), sportsmanship (accepting minor frustrations without complaint), courtesy (respecting the needs of others and behaving accordingly), and civic virtue (appropriate participation at work) to examine the presence of any OCB.

Van Dyne, Graham and Dienesch (1994) note that 'The global perception that an organization supported its members and valued their contributions was an important correlate of employee behaviour and affective states.' Eisenberger, *et al.* (2001) argue that 'based on the norm of reciprocity, employees are motivated to compensate beneficial treatment by acting in ways that support the organization'. However, they continue, 'employees may differ in their acceptance of the norm of reciprocity that underlies the exchange relationship'.

Whether employee behaviour constitutes OCB is determined by causal analysis which, in many cases, is problematic. Contractual requirements should be express, provided a contract of employment exists. Norms are established by custom and practice and may change (rapidly) in response to conditions/situations. Thus, norms of employee behaviour can be amended by pressures (threats, power exercising, inducements) by employers/bosses, resulting in apparent rather than real OCB by employees/subordinates. Hence, in practice, especially during recessionary periods and other times of difficulty in securing (alternative) employment, it can be very difficult to identify OCB.

CSR is discretionary behaviour by the organisation, in excess of the requirements of law and the norms of the market(s). CSR is evidenced most commonly by ethical behaviour towards customers, society and the natural environment. Sharp Paine (2003) documents a variety of case studies concerning potential CSR actions and demonstrates that such organisational behaviour is not only a question of actions, but that the timing and overt causes of the actions are germane.

Organisations may be tempted to use CSR actions as marketing and legal defence mechanisms in efforts to improve profitability; such motivations may be criticised from a 'pure CSR' perspective as there has been no value change in such organisations but, instead, CSR is employed as (part of) the means to the end. Instrumental values have altered, while situational values have not. This, in itself, does raise the question of whether it is the actions and their consequences that are important or the reasons for the actions (the means–ends dichotomy as in deontology–teleology). Perhaps, given the importance of environmental preservation and ethical behaviour towards others, it is appropriate for a project supply chain manager to adopt the pragmatic stance of examining behaviour and effects (the phenomenal level) in preference to the reasons underpinning the behaviour (the ideational level). However, as with all cultural aspects, if a long term (enduring) change is desired, the people concerned must be convinced of the merits of the change, so that it will become part of their usual behaviour.

3.7   Teams and Teamwork

A team is two or more people who are collaborating in pursuit of a common objective(s) – goal congruence; which distinguishes a team from a group. The people constituting a team may be quite different from each other, notably in technical knowledge and abilities, and, probably, in socio-political skills too. What is important is their preparedness and ability to collaborate towards realisation of the goal(s) which, over enduring periods, is likely to require subjugation of individual desires and behaviour for the ultimate outcome. Such collaboration is dependent upon the team members recognising their inter-dependence in striving for success and, then, acting according to that recognition (see, e.g., Crainer, 1996).

The rhetoric of teams and teamwork has been widespread and strong in the construction industry for many years – including partnering, alliances and joint venturing. Synergetic performance is believed to result from team-work. In many instances, such beliefs are coupled with competition within teams (for membership) and between teams as performance stimulants (for rewards) as well as some elements and degrees of conflict as further motiva-tors for performance (see, e.g., Robbins, 1984). However, competitive perspectives are culturally bound and, although seemingly apposite for Western participants and contexts, may be quite inappropriate elsewhere (e.g. Asia).

As teams and teamworking are dependent upon integration, communica-tion, self-subjugation and coordination (see, for example, Belbin, 1981), their existence is rare – it is exactly the lack of those key constituents for which the construction industry is criticised (see, e.g., Latham, 1994; Con-struction Industry Review Committee, 2001). As the size of any team is subject to upper limits, it is appropriate to view project TMOs as collectives of collectives; in which each collective may be a group or a team, dependent upon its constituents and processes and contingent upon its environment. Further, it is probable that if the TMO comprises groups, the TMO itself (as a meso-level collective) cannot be a team.

Here, personal dispositional variables impact the potential and probabil-ity of whether a human collective will behave as a group or as a team. Those dispositional variables are rooted in national cultures – notably, Individual-ism and Power Distance. The Femininity element of nurturing and the pursuit of Longer-Term perspectives reinforce tendencies towards teamwork by fostering integration and self-subjugation to a common good. The super-imposed factors of organisational culture and climate filter and mediate the manifestations of the basic cultural traits.

Thus, it seems that, commonly, efficient and effective supply chain func-tioning is hampered by the absence of real teams and teamwork. Whilst suitable supply chain structuring and systems may be conducive to the development of teams and teamworking, it is the forging of collaborative relationships between appropriate combinations of persons (see, e.g., Belbin, 1981, who discusses characteristic requirements for successful teams) which is critical.

3.8   (Strategic) Alliances

A business alliance is 'an ongoing, formal business relationship between two or more independent organizations to achieve common goals' (Sheth and

Parvatiyar, 1992). ul-Haq (2003) suggests that there are four principal types:

- A formal co-operative venture;
- The joint venture;
- Joint ownership;
- A strategic investment in a partner.

Whatever the formal arrangements are which bring businesses into close contact for individual transactions at one extreme or for enduring alliances at the other, 'Usually the corporate culture of the most powerful or economically successful company dominates.' (Furnham, 1997). Hence, for integration to occur successfully, whether through take-over, alliancing, merger, or forming a subsidiary joint venture organisation, not only goal congruence but also compatibility of organisational cultures is critical.

There are 'two basic organizational modes of alliance: equity joint ventures (EJVs) and non-equity joint ventures (NEJVs)' (Glaister, Husan and Buckley, 1998). That classification is supported by Pangakar and Klein (2001) who adopt the classification of equity alliances or contractual relationships.

In construction, informal alliances are common. Every project may be viewed as a joint venture due to the dependence of the output on inter-relationships between participants (interdependencies). Informal alliances constitute a hybrid in which the contract binds the participants whilst the effectiveness of the project team is determined by the quality of interpersonal relationships.

Sheth and Parvatiyar (1992) employ a two dimensional analysis – purpose (strategic/operational) and parties (competitors/non-competitors) – to examine forms, properties and characteristics of business alliances. They note that the strategic purposes of alliances are growth opportunity, strategic intent, protection against external threats, and diversification. The operational purposes, on the other hand, are resource efficiency, increased asset utilisation, enhanced core competence and a closed performance gap. A contextual factor is that an alliance form may be stipulated as the legally required structure for non-domestic organisations to operate in the location – most commonly, in less developed economies. Horizontal alliances may be made with existing competitors, potential competitors, indirect competitors and (potential) new entrants, whilst vertical alliances occur with customers, potential customers, suppliers and potential suppliers.

Contractual alliances provide much greater entry and exit flexibility for participants and at much lower cost than equity alliances however, such apparent advantage results in reluctance of the participants to make significant alliance-specific investment (Pangakar and Klein, 2001). In equity alliances, the alignment of objectives and performance incentives acts to deter participants from free-riding and from other forms of opportunistic behaviour (ibid.).

Uncertainty and trust are the two primary constructs which affect formal alliance relationships and their institutional arrangements (Sheth and Parvatiyar, 1992). Bachmann (2001) examines trust and power as means for social control within business relationships. Bachmann notes that '. . . today, trust based on individual actors' integrity can only fulfil a supplementary function, compared with trust produced by institutional arrangements.' Strong institutional arrangements are demonstrated to foster the development of trust, whilst, otherwise, business actors resort to power to safeguard their interests.

Because of the cyclical nature of the demand for property development and (to a lesser extent) construction sectors, risks are perceived to be high, given the typical returns. Hence, work allocation has a universally strong focus on cost minimisation, to the potential, virtual exclusion of other considerations. Arguably, however, design and construction work should be awarded to the parties who can provide the most suitable and reliable assurance of performance for the work (package) in question, in the context of also being compatible with parties already engaged, and others likely to be engaged over the course of the project's realisation.

Given the importance of relationships and behaviour to the operation and performance (success) of joint ventures, it is clear that culture has a fundamental impact, especially when considering compatibilities amongst participants. Those concerns are reinforced by Das and Teng (1999), who note that 'Because of incompatible organizational routines and cultures, partner firms often do not work together efficiently.'

Studies have often used measures of cultural distance, investment risk and market potential to explain which mode of entry to employ in new markets. Such decisions reflect how the firms respond to the externalities which they perceive in the target location.

> '. . . firms choose a higher control form in response to conditions of high external (market and political) uncertainty . . . [and] . . . in countries that have greater market potential. . . . firms . . . need to get established early in emerging markets . . . regardless of the market potential and/or country risk, firms resort to sharing of risks and managerial resources.' (Agarwal, 1994).

Brouthers and Brouthers (2001) investigate the relationship between cultural distance and entry mode and find that investment risk moderates that relationship.

Shenkar (2001) recognises the impact of the theory of familiarity in that firms are less likely to invest in markets which they perceive to be culturally distant. 'Follow-my-leader' strategies are often adopted by oligopolistic organisations as a method for reducing risks. Organisations which are second or later entrants to new markets are likely to adopt similar forms of entry to those adopted by initial entrants.

Kogut and Singh (1988) find that both the cultural distance from the home country and the score for Uncertainty Avoidance are correlated with preference for JV form of entry to a new country market.

Measurement of cultural distance is, itself, an issue for debate. Normally, cultural distance has been measured through use of indices (as in Kogut and Singh, 1988, who employed Hofstede's, (1980) initial four dimensions of national culture). However, that approach to measuring cultural distance involves assumptions which may be inappropriate. The problems include the fact that cultural distance is not symmetrical; home-country culture is embedded in the firm, host-country culture is embedded both in the alliance partner(s) and in the local, operating environment (Shenkar, 2001).

Measures of cultural distance as aggregate indices of measure of dimensions of culture are challenged through concerns over the relative sizes of in-group versus between-group variances. Further, cultures are dynamic temporally and vary within national borders. Not all cultural facets are of equal importance nor do they, necessarily, operate in the same direction. Intra-cultural variations (national and organisational) may exceed inter-cultural variations (Au, 2000). Hofstede (1989) confirms that differences between cultures vary in significance and that differences in Uncertainty Avoidance are, potentially, the most problematic for international business alliances. However, some emphasise Power Distance, and others focus on Individualism (versus Collectivism, in its developing context of horizontal and vertical components).

Adopting the paradigm that all construction projects are realised through (informal) joint venturing, and that culture varies to a large degree both between organisations and within societies, the issues relating to international alliances apply also to domestic construction projects.

3.9   Supply Chain Participants and Behaviour

Figure 3.1 provides a schematic, systems representation of the construction project. Not only are there extensive differences between the value perspectives of the various individual actors performing the functions identified; there will also be differing value perspectives amongst team members that constitute project actor firms. Those values may be classified as business values, technical values and personal values.

Business values, and, most particularly, the performance criteria/objectives (and parameters) derived from them, have much global commonality; however, because the performance focus is self-oriented (ultimately, at least), those values are likely to give rise to conflicts between members of the supply chain. Technical values concern the specialist activity and how it is carried out; thus, for each specialist, the values are bespoke and tend to be complementary with the technical values of other participants. Personal values are the most overtly culturally determined and variable.

Construction projects are realised through TMOs which have highly diverse members, many of whom have involvement or roles that are transient, and subject to highly varying degrees of integration. Usually, the TMOs operate, not as teams, but as flexible, multi-goal coalitions based upon fluid power structures. 'Adversarialism and opportunism are rife at all stages, as low barriers to entry maintain the high degree of fragmentation and low levels of profitability and investment within these markets'. However, '. . . construction companies are effectively the 'integrator' for a myriad of construction supply chains. . . . are faced with the challenge of obtaining a regular workload that is sufficiently profitable . . .'(Ireland, 2004).

'In general terms, it can be argued that supply chains must exist as structural properties of power . . . the physical resources that are necessary to construct a supply chain will exist in various states of contestation . . . based on the horizontal competition between those who compete to own and control a particular supply chain resource . . . also . . . on the vertical power struggle over the appropriation of value between buyers and suppliers at each point in the chain. . . . possession of these power attributes will be demonstrated by the relative capacity of the owners of particular resources to appropriate value for themselves' (Cox, 1999). Power, according to Emerson (1962), '. . . is a property of the social relation . . .' and '. . . resides implicitly in the other's dependency' and, so, is context dependent.

In construction, value measurements and perceptions are based upon comparisons of anticipations/forecasts/expectations of performance with performance realisations. A consequence is that cognitive dissonance (the mental conflict which arises when assumptions are contradicted by new information (here, when the performance realised falls significantly short of the forecasts given) (Festinger, 1957)) is likely to occur and constitute a component of the value perceived (for both the supplier and the consumer). Such comparisons, and their consequences for value perceptions, constitute a significant area of risk for project participants in addition to the tangible risks. Additionally, the vast number of interdependent, component transactions, coupled with diversity amongst participants, leads to complexity and, consequently, boundary management issues and risks.

Thus, comparisons of actual performance with forecasts and targets occurs throughout project realisation, as essential parts of performance control by managers, as well as on completion of the project supply process. Differences in performance realisations from these forecasts are, almost invariably attributed to the realisation processes and ignore the presence of (possibly considerable) variability in the forecasts themselves. Given that attention tends to focus on performance realisation shortfalls and a human tendency to blame others readily, it is all too easy for people to become frustrated and demotivated and, especially clients, generally dissatisfied with the performance achieved. The desire to avoid responsibility and consequences (liquidated damages, etc.) encourages project actors to shift blame on to others, especially those in weaker positions, as well as to pursue other elements of opportunistic behaviour – notably, claims (see Rooke, Seymour and Fellows, 2003; 2004).

Lawrence and Lorsch (1967) investigate the dichotomy of differentiation and integration within organisational processes and determine that appropriate degrees of both are required for effectiveness and efficiency – analogous to clan organisational culture. However, Tavistock (1966), and virtually all subsequent reports on construction industry organisation, criticise its performance and cite causes rooted in fragmentation, poor communications, low levels of coordination and lack of trust; indeed, the industry remains characterised by 'mutual mistrust and disrespect' amongst participants.

It is usual for fragmentation to be identified as a major cause of the problems on construction projects. Fragmentation arises through two forces – a strong force for differentiation (specialisation, division of labour, etc.)

but a weak force for integration. The result is proliferation of separate organisations which operate largely independently in pursuit of their own interests. The effects are compounded through the operation of the common procurement methods which, '. . . have focussed on organisations' individual . . . capability rather than their collective ability to integrate and work together effectively (Baiden, Price and Dainty, 2006). That is reflected in the zero-sum games typical of construction projects.

Although a vast amount of rhetoric concerns teams and teamwork in realisations of construction projects (see, e.g., Latham, 1994), in practice precious little teamwork can be found. Nicolini (2002) notes five categorical factors which are critical to success and superior performance of cross-functional teams – task design, group composition, organisational context, internal processes, and group psychosocial traits. Those factors are important contributors to 'project chemistry', which is a range of antecedent variables necessary for project management success. Dainty, *et al.* (2005) assert that project affinity, emotional attachments to the project (objectives/purpose) outcome, enhances how people work, especially their organisational citizenship behaviour (OCB), thereby fostering performance. Both constructs are culturally bound and, to their degrees of presence, enhance performance via team formation and commitment of personnel.

Given the pressures on businesses to secure competitive (financial) returns continuously, it is unsurprising that the organisational members of supply chains/networks, and their representatives on projects, succumb to opportunistic behaviour aimed at appropriation of value for self. Many systems and procedures in common usage encourage such behaviour either overtly as in 'lowest competitive bid wins' or implicitly through tight regulation as under many conditions of contract. In particular, it is such combinations which appear highly detrimental to the well-being of the industry – well-being is manifested in levels of return for all participants commensurate with the risks assumed, a cooperative and collaborative context, and levels of trust which require minimum surveillance and enforcement to secure high and continuously improving levels of performance.

Elmuti and Kathawala (2001) note that the main risks and problems which strategic alliances and, following the joint venture paradigm adopted here, construction project supply chain/network members, face are:

- 'Clash of cultures' and 'incompatible personal chemistry';
- Lack of trust;
- Lack of clear goals and objectives;
- Lack of coordination between management teams;
- Differences in operating procedures and attitudes among partners;
- Relational risk (due to self-interest focus).

There is a cultural dimension to each of the items on the list. Differences, of course, need not yield negative outcomes, indeed, positive effects of differences are emphasised in development of effective teams (as noted, above). Thus, it is not the differences themselves which are detrimental, but the ways in which they are managed – or, perhaps more appropriately, *not managed*.

Social capital is the '. . . goodwill that is engendered by the fabric of social relations and that can be mobilized to facilitate action' (Adler and Kwon, 2002). As such, social capital constitutes a powerful intangible resource with particular importance for the formation and working of an informal system of relationships amongst individuals and organisations. Social capital comprises two primary components – bridging and bonding.

'Bridging social capital examines the external linkages of individuals and groups that help to define their relationships . . . bonding social capital focuses on the internal relationships of a focal actor and specifically examines the linkages and corresponding relationships among individuals and groups within a focal group or organization' (Edelman, Bresnen, Newell, Scarborough and Swan, 2004). The two components are important in determining membership of a group or team and how the teams integrate and relate to each other and their members. However, Edelman *et al.* warn that '. . . loss of objectivity is a function of actors becoming deeply embedded in an existing network. This can lead to the exclusion of new actors or ideas that are potentially beneficial'.

Newell, Tansley and Huang (2004) examine the use of social capital for acquisition and sharing of knowledge amongst members of a project realisation collective. Social capital is important as the knowledge in question is personal/tacit knowledge and so the possessors of knowledge must become aware of its existence and then be willing and able to communicate it to relevant other parties. That exchange process is important in realisations of projects, especially for those at the forefront of knowledge and involving innovation.

Newell *et al.* (2004) note that the '. . . project team must develop "strong" relationships internally if the information and knowledge derived from . . . external networks is to be integrated'. However, ". . . individual members did appear to be using their social capital, but more for their own personal good than for the public good of the project. . . . as the project became more insecure, the individual team members increased their networking with their functional departments but very much to secure their own personal goals." Hence, trust, motivation and commitment are vital ingredients of bridging and bonding behaviour.

Goal congruence, an essential component of teamwork, arises from setting, communicating and accepting of goal(s). Further, the goals themselves may constitute performance incentives, depending on their content and their level. Goal content concerns the subject of the goal (e.g. time performance) and operates within the context of the total project – notably, the project function as in 'project affinity' as discussed by Dainty *et al.* (2005). Goal level concerns the quantity of the goal and so, its incentivisation may enhance or detract from the motivation of the goal content. Locke, Latham and Erez (1988) summarise the situation as 'commitment declines as the goal becomes more difficult and/or person's perceived chances of reaching it decline'; hence the motivation of increasingly difficult to achieve goals is an inverted U-shape. Further, they recognise that the effectiveness of different styles adopted for setting goals depends on the culture(s) (values) of the participants – perhaps, Power-Distance in particular.

Cultures evolve in path-dependent directions, punctuated by periods of stability and by rapid, step-type changes, 'The evolution of culture is shaped by agency and power, but cannot be created by fiat' (Weeks and Gulunic, 2003). However, '... despite agreement that cultural evolution occurs ... , espoused approaches to culture interventions are more commonly revolutionary in nature ...' (Harris and Ogbonna, 2002). When faced with change, most people exhibit a strong preference for the familiar and so tend to resist; if change does occur, there is a strong tendency to revert to prior norms.

Perspectives on changes in cultures span two, extremes. 'Functionalists' believe that organisational culture can be directly controlled by management and so are instrumental in promoting the cultural basis for determining organisational performance. The alternative perspective regards culture as a context within which action must be taken and so necessitates compatibility of action with the cultural environment. However, a third category, falling between these two extremes, is the perspective that culture is malleable and so may be adapted – albeit that adaptations are likely to be difficult, replete with ethical problems and require effort over long periods.

Even the most carefully devised and conducted change initiatives are likely to have unanticipated consequences – including ritualisation of change, cultural erosion, hijacking of the process, and uncontrolled and uncoordinated effects (Harris and Ogbonna, 2002).

3.10   Conclusion

Awareness, understanding and accommodation of culturally based differences are important for successful performance because success is judged by culturally determined performance metrics. Some regard culture as a tool which may be employed to effect changes to advance performance against, often pre-determined and, sometimes distantly determined, criteria and targets. This approach tends to confuse effecting cultural change (long term, permanent) with behaviour modification (short-term and reversed easily); here it is important to recall that people are risk averse and so, endeavour to return to the status quo. Others regard culture as a medium in which adaptation must occur but in which 'creep' (evolutionary change) takes place, usually steadily but with occasional perturbations, yielding long-term change.

It is inevitable that all construction projects and their realisation supply chains include a variety of cultures, whether organisational, national or both. Given this environment, successful project supply chain managers must be interculturally competent, especially to get the best from the contributing stakeholders/actors (individuals and organisations), many of whom change rapidly or end frequently throughout the life of the project. Intercultural competence requires the managers to think and act in ways which are appropriate to the cultures involved and with empathy for the various cultures, i.e. to see the project through the eyes of the different stakeholders and to appreciate the (performance) requirements which they place on the project (from their own points of view).

Part A

Such understanding and sensitivity necessitates a high level of open mindedness, flexibility and tolerance of others which is likely to be vested in only the more adventurous persons; those who regard new situations as desirable challenges, rather than threats (i.e., persons with low Uncertainty Avoidance). These attributes will be manifested in their conflict management style, which requires attention to their own and others' objectives, coupled with the ability to evaluate immediate and longer term outcomes.

It is important that managers of supply chain/network TMOs are aware of the myriad issues and difficulties concerning culture – identification, understanding, accommodation, adaptability, etc. This requires sensitivity to others as well as self-awareness. A common problem is the delusion of control – nowhere is this so evident and so important as in the selecting of appropriate combinations of people, organisations and processes to foster an environment conducive to successful performance, whatever this is determined to be for the combination of stakeholders in a construction project. Project performance results from the amalgam of the members of the supply chain, not just their 'technical' abilities, but their ability and preparedness to cooperate; only then, can synergy result.

Whilst many devote great energy to devising systems and procedures to facilitate control, much control is illusory. Good systems, including contracts, provide frameworks, and only frameworks within which projects can be realised. Selection of participant organisations can render collaborative working more (or less) likely. Such facets of 'project hardware' cannot guarantee performance or satisfaction and success, which may stem from good performance; these can be secured only through the 'project software', the people involved with the project and how they relate to the others, the organisations and the project itself.

This chapter has addressed the behavioural aspects of the 'project software', what underpins the behaviour exhibited and what the consequences are likely to be. People are fickle – clients, superiors, colleagues and subordinates – and to ignore that is to ignore an important reality. Despite the voices of critics, the overwhelming weight of evidence does portray people as knowledgeable and skilled specialists who want to do a good job. What stands in their way, *our way*, is lack of integration.

The appropriate maxim seems to be 'We're all in this TOGETHER'.

References

Adler, P.S. and Kwon, S.W. (2002) Social capital: prospects for a new concept. *Academy of Management Review*, 27(1), 17–40.

Agarwal, S. (1994) Socio-cultural distance and the choice of joint ventures: a contingency perspective. *Journal of International Marketing*, 2(2), 62–80.

Alderfer, C.P. (1972) *Existence, Relatedness and Growth: Human Needs in Organizational Settings*. New York: Free Press.

Ashforth, B.E. (1985) Climate formation: issues and extensions. *Academy of Management Review*, 10(4), 837–847.

Au, K.Y. (2000) Inter-cultural variation as another construct of international management: a study based on secondary data of 42 countries. *Journal of International Management*, 6(3), 217–238.

Bachmann, R. (2001) Trust, power, and control in trans-organizational relations. *Organization Studies*, 22(2), 337–367.

Baiden, B.K., Price, A.D.F. and Dainty, A.R.J. (2006) The extent of team integration within construction projects. *International Journal of Project Management*, 24(1), 13–23.

Belbin, R.M. (1981) *Management Teams: Why They Succeed or Fail*. Oxford: Butterworth-Heinemann.

Brouthers, K.D. and Brouthers, L.E. (2001) Explaining the National Cultural Distance paradox. *Journal of International Business Studies*, 32(1), 177–189.

Burns, T. and Stalker, G.M. (1961) *The Management of Innovation*. Tavistock, 2nd ed., 1968.

Cameron, K.S. and Quinn, R.E. (1999) *Diagnosing and Changing Organizational Culture*. Massachusetts: Addison-Wesley Longman.

Chen, C.C., Meindl, J.R. and Hunt, R.G. (1997) Testing the effects of vertical and horizontal collectivism: a study of allocation preferences in China. *Journal of Cross-Cultural Psychology*, 28(1), 44–70.

Construction Industry Review Committee (2001) *Construct for Excellence* ('The Tang Report'). Hong Kong: Government of the Hong Kong Special Administrative Region.

Cox, A. (1999) Power, value and supply chain management, *Supply Chain Management*, 4(4), 167–175.

Coyle-Shapiro, J.A-M. (2002) A psychological perspective on organizational citizenship behavior. *Journal of Organizational Behavior*, 23(8), 927–946.

Crainer, S. (1996) *Key management ideas: thinkers that changed the management world*. London: Pitman Publishing.

Dainty, A.R.J., Bryman, A., Price, A.D.F., Greasley, K., Soetanto, R. and King, N. (2005) Project affinity: the role of emotional attachments in construction projects. *Construction Management and Economics*, 23(3), 241–244.

Das T.K. and Teng, B-S. (1999) Managing risks in strategic alliances. *The Academy of Management Executive*, 13(4), 50–62.

Denison, D. (1996) What is the difference between organizational culture and organizational climate? A native's point of view on a decade or paradigm wars. *Academy of Management Review*, 21(3), 619–654.

Edelman,L.F., Bresnen, M., Newell, S., Scarborough, H. and Swan, J. (2004) The Benefits and Pitfalls of Social Capital: Empirical Evidence from Two Organizations in the United Kingdom. *British Journal of Management*, 15(S1), 59–69.

Eisenberger, R., Armeli, S., Rexwinkel, B., Lynch, P.D., and Rhoades, L. (2001) Reciprocation of perceived organizational support, *Journal of Applied Psychology*, 86(1), 42–51.

Elmuthi, D. and Kathawala, Y. (2001) An overview of strategic alliances. *Management Decision*, 39(3), 205–217.

Emerson, R.M. (1962) Power-Dependence Relations. *American Sociological Review*, 27(1), 31–41.

Fellows, R.F. (2006) Understanding approaches to culture. *Construction Information Quarterly*. 8(4), 159–166.

Fellows, R.F., Liu, A.M.M. and Storey, C. (2004) Ethics in construction project briefing, *Journal of Science and Engineering Ethics*, 10(2), 289–302.

Festinger, L. (1957) *A Theory of Cognitive Dissonance*. Stamford: Stamford University Press.

Part A

Furnham, A. (1997) *The psychology of behaviour at work: the individual in the organization*. Hove: Psychology Press.

Glaister, K.W., Husan, R. and Buckley, P.J. (1998) UK International Joint Ventures with the Triad: Evidence for the 1990s. *British Journal of Management*, 9(3), 169–180.

Gomez, C., Kirkman, B.L. and Shapiro, D.L. (2000) The impact of collectivism and in-group / out-group membership on the generosity of team members. *Academy of Management Journal*, 43(6), 1097–1100.

Green, S.D. and Simister, S.J. (1999) Modelling client business processes as an aid to strategic briefing. *Construction Management and Economics*, 17(1), 63–76.

Griffith, A. and Sidwell, A.C. (1995) *Constructability in Building and Engineering Projects (Building & Surveying.*, Basingstoke, Macmillan.

Handy, C.B. (1985) *Understanding Organisations (3 edn.)*. Harmondsworth, Penguin.

Harris, L.C. and Ogbonna, E. (2002) The unintended consequences of cultural interventions: a study of unexpected outcomes. *British Journal of Management*, 13(1), 31–49.

Hatch M.J. (1993) The dynamics of organisational culture. *Academy of Management Review*, 18(4), 657–693.

Herzberg, F., Mausner, B. and Bloch Snyderman, B. (1967) *The motivation to work (2 edn.)* New York: Wiley.

Hinman, L.M. (1997) *Ethics: A Pluralistic Approach to Moral Theory*. Orlando: Harcourt Brace Jovanovich.

Hofstede, G.H. (1980) *Culture's consequences: international differences in work-related values*. Beverley Hills, CA.: Sage Publications.

Hofstede, G.H. (1983) The cultural relativity of organizational practices and theories. *Journal of International Business Studies*, 14(Fall), 75–89.

Hofstede, G.H. (1989) Organizing for cultural diversity. *European Management Journal*, 7(4), 390–397.

Hofstede, G.H. (1994a) The business of international business is culture. *International Business Review*, 3(1), 1–14.

Hofstede, G.H. (1994b) *Cultures and organizations: software of the mind.*, London: Harper Collins.

Hofstede, G.H. (2001) *Culture's Consequences: Comparing Values, Behaviors, Institutions, and Organizations Across Nations (2 edn.)*. Thousand Oaks, CA: Sage.

Hutton, W (1996) *The State We're In (2 edn.)*. London: Vintage.

Hutton, W. (2002) *The World We're In*. London: Abacus.

Ireland, P. (2004) Managing appropriately in construction power regimes: understanding the impact of regularity in the project environment. *Supply Chain Management*, 9(5), 372–382.

Jin, X-H, Doloi, H. and Goa S-Y. (2007) Relationship-based determinants of building project performance in China. *Construction Management and Economics*, 25(3), 297–304.

Kogut, B. and Singh, H. (1988) The effect of national culture on the choice of entry mode, *Journal of International Business Studies*, 19(3), 411–433.

Kotler, P. (1972) A Generic Concept of Marketing. *Journal of Marketing*, 36(2), 46–54.

Kroeber, A.L. and Kluckhohn, C. (1952) Culture: a critical review of concepts and definitions. In *Papers of the Peabody Museum of American Archaeology and Ethnology*, Vol. 47, Cambridge, MA.: Harvard University Press.

Latham, Sir M., (1994) *Constructing the Team*. London: HMSO.

Lawrence, P.R. and Lorsch, J.W. (1967) *Organization and environment: managing differentiation and integration*. Boston: Division of Research, Graduate School of Business Administration, Harvard University.

Leary, M.R. (1991) *Introduction to behavioral research methods*. Belmont, CA.: Wadworth.

Locke, E.A., Latham, G.P. and Erez, M. (1988) The determinants of Goal Commitment. *The Academy of Management Review*, 13(1), 23–39.

Maslow, A.H. (1943) A Theory of Human Motivation. *Psychological Review*, 50(4) July, 370–396.

Mills, G., Austin, S., and Thomson, D. (2006) Values and value: two perspectives on understanding stakeholders. *Proceedings, Joint International Conference on Construction Culture, Innovation and Management (CCIM)*, The British University in Dubai, November, CD-Rom.

Mullins, L.J. (2002) *Management and Organisational Behaviour* (6 edn). Harlow: Prentice Hall.

Newell, S., Tansley, C. and Huang, J. (2004) Social capital and knowledge integration in an ERP project team: The Importance of Bridging AND Bonding. *British Journal of Management*, 15(S1), 43–57.

Nicolini, D. (2002) In search of 'project chemistry'. *Construction Management and Economics*, 20(2), 167–177.

Organ, D.W. (1988) *Organizational Citizenship Behavior: The Good Soldier Syndrome*. Lexington MA.: Lexington Books.

Pangakar, N. and Klein, S. (2001) The Impacts of Alliance Purpose and Partner Similarity on Alliance Governance. *British Journal of Management*, 12(4), 341–353.

Robbins, S.P. (1984) *Essentials of Organizational Behaviour*. Englewood Cliffs, NJ.: Prentice-Hall.

Rokeach, M. (1973) *The Nature of Human Values*. New York: Free Press.

Rooke, J., Seymore, D.E. and Fellows, R.F. (2003) The Claims Culture: A Taxonomy of Attitudes in the Industry. *Construction Management and Economics*, 21(2), 167–174.

Rooke, J., Seymore, D.E. and Fellows, R.F. (2004) Planning for claims: an ethnography of industry culture. *Construction Management and Economics*, 22(6), 655–662.

Rosenthal, R. and Rosnow, R.L. (1991) *Essentials of Behavioral Research: methods and data analysis (2 edn.)*. Boston, Mass.: McGraw-Hill.

Schein, E.H. (1984) Coming to an Awareness of Organizational Culture, *Sloan Management Review*, 25(Winter), 3–16.

Schein, E.H. (1990) Organisational Culture. *American Psychologist*, 45, 109–119.

Schneider, B. (1975) Organizational climate: an essay. *Personnel Psychology*, 28, 447–79.

Schneider, W.E. (2000) Why good management ideas fail: the neglected power of organizational culture. *Strategy and Leadership*, 28(1), 24–29.

Schwartz, S.H. and Bardi, A. (2001) Value hierarchies across cultures: taking a similarities perspective. *Journal of Cross-Cultural Psychology*, 32(3), 268–290.

Schwartz, S.H. and Bilsky, W. (1987) Toward a psychological structure of human values. *Journal of Personality and Social Psychology*, 53(3), 550–562.

Schwartz, S.H. and Sagiv, L. (1995) Identifying culture-specifics in the content and structure of values. *Journal of Cross-Cultural Psychology*, 26(1), 92–116.

Sharp Paine, L. (2003) *Value Shift: Why Companies Must Merge Social and Financial Imperatives to Achieve Superior Performance*. New York: McGraw-Hill.

Part A

Shenkar, O. (2001) Cultural distance revisited: towards a more rigorous conceptualization and measurement of cultural differences. *Journal of International Business Studies*, 32(3), 519–535.

Sheth, J.N. and Parvatiyar, A. (1992) Towards a theory of Business Alliance Formation. *Scandinavian International Business Review*, 1(3), 71–87.

Tagiuri, R. and Litwin, G.H. (Eds.) (1968) *Organizational Climate*. Graduate School of Business Administration, Harvard University.

Tavistock Institute of Human Relations (1966) *Interdependence and Uncertainty: a study of the building Industry*. London: Tavistock Publications.

The Chinese Culture Connection (a team of 24 researchers) (1987) Chinese values and the search for culture-free dimensions of culture. *Journal of Cross-Cultural Psychology*, 18(2), 143–164.

Thorelli, H.B. (1986) Networks: between markets and hierarchies. *Strategic Management Journal*, 7(1), 37–51.

Triandis, H.C. (1990) Cross-cultural studies of individualism and collectivism. In Berman, J.J. (Ed.), *Cross-cultural perspectives, Nebraska Symposium on Motivation 1989*, Lincoln: University of Nebraska Press, 41–133.

Trompenaars, F. and Hampden-Turner, C. (1997) *Riding the Waves of Culture* (2 Edn.). London: Nicholas Brealey.

ul-Haq, R. (2003) *Executive Briefing – Strategic Alliances*. Birmingham: the Birmingham Business School.

Van Dyne, L., Graham, J.W., and Dienesch, R.M. (1994) Organizational citizenship behaviour: construct redefinition, measurement and validation. *Academy of Management Journal*, 4(4), 765–802.

Victor, B. and Cullen, J.B. (1988) The organizational bases of ethical work climate. *Administrative Science Quarterly*, 33(1), 101–125.

Vroom, V. (1964) *Work and Motivation*. New York: Wiley.

Wayne, S.J., Shore, L.M. and Linden, R.C. (1997) Perceived organizational support and leader-member exchanges: a social exchange analysis. *Academy of Management Journal*, 40(1), 82–111.

Weeks, J. and Gulunic, C. (2003) A theory of the cultural evolution of the firm: the intra-organizational ecology of memes, *Organization Science*, 24(8), 1309–1352.

Williams, A., Dobson, P. and Walters, M. (1989) *Changing culture: new organizational approaches*. London: Institute of Personnel Management.

Learning to Co-operate and Co-operating to Learn: Knowledge, Learning and Innovation in Construction Supply Chains

Mike Bresnen

Although construction is a setting where contracting has traditionally been seen as adversarial, adding further to the problems of promoting learning and innovation across the industry (e.g. Latham, 1994), recent developments in supply chain management (SCM) – including the advent of 'partnering' and 'alliancing' – suggest the development of more cooperative, long-term relationships between clients, contractors and subcontractors that are on the face of it more conducive to the spread of knowledge, learning and innovation (Egan, 1998). Such initiatives promise not only improvements in project performance, but also greater responsiveness to client needs and improved innovation potential (Bennett and Jayes, 1995).

Despite the growth of interest in collaboration in supply chains, there remain however many unanswered questions about the effects on knowledge, learning and innovation processes. For example, how is relevant knowledge and expertise diffused across the highly complex and fragmented division of labour found in construction supply chains? How are (inter-) organisational cultures within supply chains shaped to support learning and knowledge creation? How is knowledge and experience of new techniques and processes externalised/articulated and codified (and with what effects)? How important are social processes in the diffusion of knowledge and learning and how do the practicalities of supply chain interaction affect these processes?

Given the importance of communities or networks of practice to processes of learning (Brown and Duguid, 2001), the novelty and idiosyncratic nature of project-based work poses significant problems and constraints when it comes to attempting to capture, diffuse and embed knowledge and learning about new technologies, systems of organisation and ways of working (Bresnen, 2006). Not only may each project be very different, but knowledge and expertise may also be lost as project teams disband and move on to new projects. Where learning about new ways of working depends upon the tacit knowledge of individuals and the collective experience of teams, valuable lessons learned may be lost and attempts to capture and codify knowledge in new systems and procedures may be thwarted.

In an attempt to throw some light on the questions above, this chapter draws upon findings from research conducted by the author, which explores collaborative relationships within the construction industry setting, investigating in particular the practice of supply chain partnering between clients, contractors and subcontractors (and using illustrative quotes from research reported in Bresnen and Marshall (2000)). The chapter draws upon social- or process-based perspectives on knowledge, innovation and learning that emphasise the situated and socially embedded nature of learning within and across organisations (e.g. Newell *et al.*, 2002). It further examines SCM not only as a vehicle for knowledge, learning and innovation within the industry, but also – in its impact as new management practice – as an *object* of knowledge sharing, learning and innovation within the industry. The intention here is to understand not only how collaboration affects knowledge sharing, learning and innovation processes within construction supply chains, but also how construction supply chains themselves facilitate and inhibit the spread of new management practices, such as partnering, that ostensibly require 'cultural change' within the industry (Egan, 1998).

4.2   Supply Chain Management: Innovation, Knowledge Sharing and Organisational Learning

Across a wide range of industrial sectors, there has been a trend over the last twenty years away from the traditional 'arms length' or adversarial relationships that have often characterised transactions between contractors and their suppliers, towards various forms of 'obligational' or 'relational' contracting that involve the establishment of more cooperative forms of relationship (e.g. Dwyer *et al.*, 1987; Sako, 1992). Often taking the form of strategic partnerships or alliances, such changes in practice have occurred in tandem with the outsourcing of business functions and attempts to rationalise supply bases. From an operations management and purchasing perspective, such developments have crystallised around the concept of SCM (Christopher, 1992; Harland, 1996; Lamming, 1996). This has led, in turn, to a good deal of work that has attempted to explore the factors enabling and inhibiting the management of supply chains and the systems and techniques available for organising and managing transactions – including those based on the use of information technology (e.g. Harland *et al.*, 2004).

Definitions of SCM tend to emphasise the importance of management being proactive in integrating activities and business processes across the supply chain in the interests of responding to the needs of customers and with the aim of improving task and business performance. So, for example, Cox (1999: 1) defines SCM as:

'*A way of thinking that is devoted to discovering tools and techniques that provide for increased operational effectiveness and efficiency throughout the delivery channels that must be created internally and externally to support and supply existing corporate product and service offerings to customers.*'

Beyond this, however, approaches to SCM differ considerably in practice and a good deal of attention has been directed towards identifying and exploring the range of approaches adopted. Taxonomies that have attempted to capture this variation in practice point, in particular, to the extent of backward integration (with suppliers) and/or forward integration (with customers) as marking differences in the scope and extent of integration (Bessant *et al.*, 2003; Harland *et al.*, 2001). There is also a good deal of attention directed towards understanding variation in the depth of collaboration within supply chains. At one extreme are more superficial modes of interaction, usually based on commitment to a common task and involving project management techniques and shared information systems (Fawcett and Magnan, 2002). At the other extreme, are more profoundly collaborative modes of interaction within supply chains that involve commitments to integration and resource sharing at all levels, with a longer term strategic perspective being taken on the development of the relationship (e.g. Håkansson and Snehota, 1995).

Although there is a good deal of interest in understanding and promoting the development of more fully integrated and functioning supply chain relationships – especially in sectors such as the automotive industry (Lamming, 1993) – research within the operations management and purchasing field has tended to be focused mainly on immediate, short-term transactions. This has been driven primarily by the concern to improve logistics and operational efficiency, rather than being concerned with the longer term, more 'transformational' aspects of SCM, associated with processes of innovation, knowledge sharing and learning and the more strategic development of relationships within supply networks (Harland *et al*, 2004). Theory and research on the management of supply chain relationships has certainly broadened out from an earlier, narrower concern with economic efficiency and financial performance to encompass a wider range of competitive capabilities, including product and customer focus, innovation in product and process development and inter-organizational knowledge sharing and learning. However, there is still the tendency for more operational, as opposed to strategic, perspectives to dominate operations management discourse (Green *et al.*, 2005).

Moreover, there is every reason to believe that these broader effects are important and worthy of considerably more attention than they receive. As Harland *et al.* (1999) point out, SCM is not just about improving efficiency, but also about adding value across the entire supply chain. Importantly, such value is added not just through minimising waste and improving efficiency, but also, in the longer term, through innovation and the creation, capture and exploitation of knowledge and learning – particularly in sectors (such as advanced engineering) where innovation, knowledge creation and learning are crucial to short and long term competitive advantage. Indeed, research on SCM has increasingly emphasised the potential for learning and innovation in supply relationships (Lamming, 1993; Dyer and Nobeoka, 2000), as too has work on inter-organisational networks in fields such as biotechnology (e.g. Powell *et al.*, 1996).

At the same time, however, it is clear that this potential is far from being realised. Research has tended to show that SCM programmes generally do not incorporate supply chain learning and that, where such learning does occur, it tends to be very informal, unstructured and limited to first-tier suppliers. So, for example, Bessant *et al.* (2003) set out to explore the use of supply chains as a mechanism for transferring 'appropriate practice'. In a study of supply chains ranging across several industrial sectors, they identify a number of 'blocks' to supply chain learning, which not only concern technical and logistical problems, but which also relate to organisational factors. Among the latter, they highlight the importance of internal structural divisions and 'cultural differences' that work against the adoption of a coherent approach to learning through supply chains and which affect the motivation and/or ability of those involved to engage with suppliers/customers. They also highlight issues such as lack of trust, as having a more direct negative impact on interaction.

These findings reinforce other research on SCM that has tended to find that the rhetoric of supply chain collaboration and integration is often at odds with the reality. Purchasing managers themselves often see SCM as simply the latest in a long line of managerial fads and fashions, one of the results of which is that many companies take only a very simplistic approach to SCM, relying heavily on integration through systems based on information and communication technology (Fawcett and Magnan, 2002). More sophisticated approaches that recognise the importance of relational capabilities and which develop a SCM 'culture' that permeates decision-making, are extremely rare (Kotabe *et al.*, 2002; McIvor and McHugh, 2000).

Research has also tended to emphasise that, even when companies have the willingness and ability to develop their approach to managing supply chains, the reality faced by subcontractors and suppliers further down the line is often far from the idyllic picture of collaboration and mutual benefit commonly expressed in so much of the prescriptive literature (Bresnen, 2007). The problems faced by small or medium-sized firms can be particularly acute, especially where the management of the relationship involves the relentless application of performance improvement programmes (such as continuous improvement).

Of course, subcontractors and suppliers are not always small firms and, what's more, any counter-veiling commercial power or expert knowledge they may have means that circumstances do not always necessarily work in favour of the dominant client or buyer (Bresnen, 1996). Such conditions, for example, are found in the biotechnology industry, where the research capabilities of large pharmaceutical companies can be heavily dependent upon the expertise provided by smaller biotechnology companies (Swan *et al.*, 2007). However, the more common scenario is that full supply chain integration is limited due, in part, to imbalances of power within the supply chain (Cox, 1999). This, in turn, means that the potential for collaboration and mutual benefit is conditional upon, amongst other things, the balance of power and influence between parties to the transaction (Cox and Ireland, 2002; Cox *et al.*, 2001).

4.3   SCM in the Construction Industry

Paralleling these developments in mainstream management theory and practice has been the emergence, over the last ten years or so, of various forms of relational contracting in the construction industry, including the application of SCM thinking. In the case of construction, an important part of the drive for improved supply chain interaction has been the explosion of interest, post-Latham and post-Egan, in the potential of partnering for improving collaboration between clients, contractors and subcontractors and, hence, project and company performance (e.g. Bennett and Jayes, 1995; Barlow *et al.*, 1997; Holti and Standing, 1996).

Evidence on the value and benefits of partnering per se is still mixed (Wood and Ellis, 2005) and there remain issues and debates surrounding the application of the logic of strategic partnering to a context in which inter-firm relationships do not normally extend into the long term and usually centre upon single projects or one-off programmes of activity (Bresnen and Marshall, 2000; Bresnen, 2007; Phua, 2006). Nevertheless, interest in how partnering can improve collaboration between contractual partners and so boost performance has led to a considerable amount of work aimed at understanding the nature of partnering in practice (e.g. Nystrom, 2005), as well as the economic and institutional factors facilitating and inhibiting collaboration between contractual partners (e.g. Phua, 2006). It has also dovetailed closely with work concerned more specifically with the effects of re-configuring supply chain relationships in the sector into 'clusters' in order to improve project performance (e.g. Nicolini *et al.*, 2001). There is a view that collaborative relationships that go unmanaged result in rising costs to clients and that SCM is the means through which improvements in value and reductions in costs can be achieved.

Attempts have also been made to apply directly the concept of SCM to the construction industry and to assess the extent and depth of its adoption (e.g. Briscoe *et al.*, 2004; Holti *et al.*, 2000; Saad *et al.*, 2002; Vrijhoef and Koskela, 2000). Both theory and practice are less well developed in construction when compared to other industries and there are constraints within the industry that are a factor to consider here. Nevertheless, many see considerable potential in applying SCM thinking to construction in order to improve project and/or company performance – particularly in parts of the industry, such as house-building, where project work is perhaps less idiosyncratic and more routine (Childerhouse *et al.*, 2003; Naim and Barlow, 2003). Others suggest that the longer-term relationships that tend to be associated with newer forms of procurement, such as PFI and prime contracting, may provide a longer lasting context for the embedding of SCM practices (Briscoe and Dainty, 2005; Green *et al.*, 2005).

However, the overwhelming picture that emerges from research to date paints a much less sanguine picture of the extent of adoption of SCM in the construction industry and of its effectiveness in practice (e.g. Akintoye *et al.*, 2000; Briscoe and Dainty, 2005). In a study of cases of attempted supply chain integration, Briscoe and Dainty (2005: 323), for example, identify many continuing problems in areas such as managing information and communication, aligning management systems and in achieving and improving quality standards, concluding that 'none of the clients had managed to align its supply chain partners' practices in such a way as to obtain the full benefits from supply chain integration.' Saad *et al.* (2002) depict SCM as a sophisticated 'fifth generation innovation' (Rothwell, 1994) that requires high levels of integration at both intra- and inter-organisational levels and a long-term perspective on improving business performance. However, they then go on to conclude from their survey results that there is a fundamental lack of preparedness of the industry to adopt SCM and a lack of understanding of the concept and the prerequisites needed for successful implementation.

Existing research has also identified many of the same sorts of gaps between the rhetoric and reality of SCM encountered in other industrial settings. So, for example, it is clear that SCM, like partnering, tends in practice to be restricted to 'first tier' suppliers (i.e. main contractors) rather then extending further down the supply chain to subcontractors or suppliers (Akintoye *et al.*, 2000). Other research has noted further how coordination tends to be limited and dependent on the use of project management techniques and the alignment of ICT systems, rather than involving a deeper integration of systems, practices and processes (Briscoe and Dainty, 2005). Perhaps most significant, however, is the research evidence which suggests that relationships between contractors and subcontractors are seldom considered fair or of obvious mutual benefit to both parties (Briscoe *et al.*, 2001; Dainty *et al.*, 2001; Thorpe *et al.*, 2003). Such findings add to those found from research on partnering, which suggest that the benefits of collaboration between client and contractor at the top of the supply chain is sometimes achieved at the cost of those lower down the supply chain – namely, subcontractors and suppliers (Bresnen and Marshall, 2000).

Consequently, it is important here again to understand how SCM is affected by the balance of interests and influence between organisations involved in commercial transactions (Bresnen 2007; Green *et al.*, 2005). Moreover, this is not only affected by the amount of commercial power and expertise brought by each party to the transaction (Cox *et al.*, 2001), but also by deeper underlying systems of rules and norms that govern interaction in an inter-organisational setting and which therefore may effectively privilege one group of interests or needs over another (cf. Elg and Johannson, 1997; Hardy and Phillips, 1998), as well as have different implications for innovation and the sharing of knowledge and learning between organisations (Hardy *et al.*, 2003). As the Project Manager of one mechanical and electrical services subcontractor put it:

'I've been involved in three partnering projects now and they're all the same – it's driven around cost. 'How can we get the job cheaper?' There are certain things

that we introduced here and this job will be cheaper. The next job won't, because that will be written into the spec . . . Next time, everybody will be pricing for that, so effectively all you've done is you've taken a little bit of value off the project . . . The next time, you've still got the same pressure on you – 'how are you going to get it cheaper?'

Beyond conditions of power on particular projects or programmes of work, however, explanations for the seeming inability to develop and adapt SCM thinking to the construction industry context are inevitably sought in the particular structure and 'culture' of the industry (Saad *et al.*, 2002). As Briscoe and Dainty (2005: 325) put it, 'it may well be the case that an industry that is characterised by one-off projects, wide geographical dispersal, many small firms and cyclical demand for its products and services may never be able to realise the full fruits of supply chain integration' (Briscoe and Dainty, 2005: 325). That may be so. However, such a conclusion presents a rather bleak prospect for attempts to manage supply chains and also creates a somewhat deterministic picture of the effects of the structure/culture of the industry on action.

A more telling analysis perhaps is presented by Green *et al.* (2005) who, in contrasting SCM in the construction and aerospace sectors, explore the different 'sense-making' logics applied by actors in each sector to explain their need to develop supply chain collaboration and their attempts at rationalising success or failure. Green *et al.* (2005) demonstrate that, whereas in the aerospace context there is a clear aspiration to engage with supply chain integration because it makes 'good business sense' and is in alignment with sector requirements and dynamics, in the case of construction, the approach is very different. Here, aspirations are less 'strategic' and more 'operational' – manifested in more of a concern to improve individual project performance (especially costs) than to put emphasis on longer term commercial success. Moreover, they find that attitudes are dominated by a concern with barriers to implementing SCM and with the extent of 'cultural change' supposedly needed in a context where adversarial attitudes are considered well entrenched.

Consequently, although structural/cultural conditions are important, it is not simply those conditions themselves that determine or condition future action, but the ways in which they are interpreted, enacted and reproduced that constrain opportunities for change. A similar point is made by Dubois and Gadde (2000), who suggest that fragmentation in the industry due to an emphasis on transactional, rather than relational exchange, is not inevitable, as it depends considerably upon purchasing behaviour. The points made above about power similarly emphasise that resources and sources of legitimisation *can* be mobilised by parties to a transaction to achieve their interests – without guaranteeing that they *will* be (Hardy and Phillips, 1998).

Whatever the position taken about the relationship between structure/culture and action, such analyses make it clear that it is important to emphasise again that understanding SCM depends a lot upon an understanding of context (including conditions of power) and the ways in which context may shape behaviour (without determining it) through the reproduction or alteration of routines and practices. The importance of context and the effects of one-off projects and project-based organisation on innovation, knowledge integration and learning processes associated with SCM are returned to again later. In the meantime, the discussion turns next to examining what the literature has to say directly about the effects of SCM and partnering on processes of innovation, knowledge integration and learning in the longer term.

4.4   Innovation, Knowledge Sharing and Learning in Construction Supply Chains

One of the key points to emerge from the above discussion is the pre-eminence attached to *project* performance as the main driver for interest in SCM. Where collaboration within supply chains has been shown to lead to performance gains, it is clear that an emphasis tends to get placed upon immediate and quantitative project performance (particularly time and cost reductions). Much existing work on SCM in the construction context therefore mirrors that found in other sectors, focusing on its effects on improving project performance and, notably, in reducing construction costs (Proverbs and Holt, 2000). Despite the recognition that reduced costs might be achieved through continuous improvement over the medium term and the occasional foray into other types of project outcome – such as environmental impact (e.g. Ofori, 2000) – this emphasis on the impact of SCM upon project objectives and project performance remains paramount.

In contrast, there is a comparative lack of emphasis placed upon understanding the effects of SCM upon more qualitative, intangible and longer term performance criteria – such as knowledge sharing, innovation and organisational learning. Holti and Whittle (1998) do explore learning networks in the construction industry, differentiating between 'operational learning' and 'strategic learning'. Others too have started to explore the organisational learning capabilities of project-based organisations themselves (Styhre *et al.*, 2004; Knauseder *et al.*, 2007). However, these excursions into understanding longer term processes of innovation, knowledge sharing and learning within construction supply chains are still comparatively rare.

The reason for this is undoubtedly, as Dubois and Gadde (2000) suggest, due to the strong emphasis on transactional exchange – reflected in competitive tendering for one-off projects and a focus on efficiency improvement – that not only fails to capture the benefits of the 'massive interaction' that occurs amongst actors involved in a project, but also hampers the development of wider network learning effects. Differentiating between projects as temporary networks within more permanent organisational networks, Dubois and Gadde (2000) nevertheless remain optimistic about the prospects for innovation within supply networks, arguing for increased reliance on relational exchange in order to enhance conditions for learning and adaptations amongst firms in the permanent network.

However, the evidence regarding the effects of collaboration within supply chains on knowledge and learning outcomes is, perhaps not surprisingly, mixed and unconvincing. Quite apart from continuing problems associated with managing the flow of information and communications between supply chain partners on specific projects (Bresnen and Marshall, 2000; Briscoe and Dainty, 2005), there continue to be many problems experienced in achieving the benefits of better integration in terms of value engineering, contractor input into design, quality improvements during construction and the like (Briscoe and Dainty, 2005). With regard to longer term outcomes, the evidence is even less compelling. In a survey of 118 firms in construction, for example, Saad *et al.* (2002) note a strong value being placed on the importance of sharing learning and innovation, combined, however, with a reliance on traditional mechanisms – such as in-house training and external workshops – that have limited ability to promote shared learning. Recent research in Sweden further emphasises the under developed nature of inter-organisational learning capabilities in construction, highlighting the importance of informal networking and limits to experimentation and organising as modes of learning (Styhre *et al.*, 2004).

The problem is partly due to the situated nature of learning and the difficulty in promoting and capturing innovation and learning in a context in which work is project-based (Winch, 1998; Sydow *et al.*, 2004). This issue is returned to and discussed further below. However, it is also due to related inherent antagonisms and contradictions between performance criteria themselves – an issue that is rarely addressed in the literature. It has, of course, long been noted that there are important potential antagonisms, as well as complementarities, between basic project performance criteria (time, cost, quality). So, for example, although fast completion might require more resources and so increase project costs, it is also possible that it can reduce the opportunity costs of having resources tied up and so lead to financial savings. Similarly, improved quality may or may not involve additional time and cost, depending, for example, on whether it involves the use of standardised components or bespoke work.

The same logic, however, also applies to inter-relationships amongst the various critical intervening processes leading to improvements in project performance (e.g. better contractor input into design, improved design-construct coordination, etc), as well as between these and other performance criteria at different levels of analysis. So, for example, if one takes the criteria of improved design-construct coordination, this may clearly contribute to reducing the time and cost spent on the project. However, its relationship with quality as an objective is ambiguous, as it depends on a number of other factors (e.g. whether it leads to more or less creative design elements). Furthermore, there is no necessarily symbiotic relationship between achieving better design-construct coordination and other intervening processes affecting performance (such as value engineering or quality improvement).

In fact, these may be less likely to occur if the time and effort spent upon them is considered less beneficial than the time and/or cost benefits of improved coordination.

What is more, at the level of more general company/supply chain/network performance improvement, to what extent is better design-construct coordination necessarily consistent with greater innovation, for example? The latter may require rather more in the way of 'creative abrasion' between participants than is suggested by the prospect of improved coordination (cf. Leonard-Barton, 1995). Improved coordination may certainly be associated with better systems integration between participants (Briscoe and Dainty, 2005). However, it is not necessarily consistent with longer term organisational learning, given the tendencies for inter-organisational processes to ossify into more established routines (Holmqvist, 2003). The problem here is that the tendency in the literature is to treat such performance criteria as essentially additive, whereas in fact they interconnect in much more complex ways. Moreover, they operate at different levels of analysis, so that performance improvement at one level (e.g. the project) is by no means necessarily consistent with performance improvement at another (e.g. the business relationship). Consider, for example, the following account given by the Property Development Director of a group of companies in the hotel and leisure business:

> 'We believe that we should incentivise the relationship between the parties and it should be based upon business performance and, if you like, the bigger relationship . . . That doesn't mean to say that we shouldn't have project bonuses, but I think project bonuses should be simple and should be about finishing on time [and] within budget and everyone should be on much the same sort of arrangement. I think that, for the relationship as a whole, it ought to be based on the business plan.'

A useful insight can be gained here from the application of well established ideas from organisational learning theory. Levinthal and March (1993), for example, differentiate between exploitative and explorative learning, the former involving the application of existing knowledge and learning to the more efficient and effective performance of tasks, the latter being more exploratory and concerned with the creation of new knowledge and learning. The point here is simply that a focus on project performance improvement tends to correspond to a more exploitative and localised approach to knowledge and learning (Levinthal and March, 1993). This is not to suggest that there is no possibility of innovation and learning occurring within the parameters set for the project, nor that learning cannot occur either through or between projects (although, as will be seen, this can be difficult). However, it does mean that the pursuit of project performance enhancement, on the one hand, and wider innovation and organisational learning within project-based organisational forms, on the other, can operate according to quite different logics that are not necessarily consistent – the former being more exploitative, the latter more explorative (Holmqvist, 2003).

4.5   Situated Learning and Implications for Project-Based Supply Chains

To return to the importance of context, there are a number of implications that also arise when one considers the direct effects of the project-based organisational setting within which projects are undertaken on the basis of innovation, knowledge and organisational learning processes across supply chains within the industry. The remaining discussion focuses on these issues, concentrating on organisational and social processes associated with innovation, knowledge and learning in supply chains.

First of all, research on knowledge sharing and learning has increasingly emphasised the importance of social networks and social interaction for understanding flows of knowledge and learning both within and between organisations (e.g. Brown and Duguid, 2001; Lave and Wenger, 1991). Research on project settings suggests that social interaction is no less important there for understanding the diffusion of knowledge (Bresnen *et al.*, 2003; DeFillippi and Arthur, 1998; Hansen, 2002). However, the key point about projects and project-based organisation is that, in a number of important ways, they create serious problems for the development and consolidation of the types of social relations that have been shown to enable the diffusion of knowledge and learning. The author's research into partnering, for example, has highlighted how the diffusion of knowledge and experience of inter-firm partnering is not only crucially dependent upon, but also constrained by, staff availability and associated secondment practices (Bresnen and Marshall, 2000, 2002). More generally, there is the obvious difficulty of being able to form and sustain knowledge sharing entities, such as 'communities of practice' (Brown and Duguid, 2001; Lave and Wenger, 1991) in conditions characterised by discontinuities in staffing and constant team building and rebuilding focused around specific project tasks (see also Bresnen, 2003).

From a supply chain management perspective, such regular and frequent breaks in working relationships within and between teams and their organisations obviously create serious constraints upon developing the sorts of continuing social interaction necessary to share and exploit knowledge and learning over the long term. Moreover, researchers such as Lindkvist (2005) have gone further, by suggesting that projects teams, in acting as knowledge 'collectivities' (as opposed to 'communities'), are even less likely to engage in the sharing of their specialised knowledge to create new knowledge. If this is the case, it becomes even more difficult to imagine the likelihood of engaging in more explorative modes of learning between organisations involved in supply chain relationships. Other factors, such as the possibility of staff working in highly competitive internal organisational environments or the importance of professional indemnity requirements, may also clearly have an influence.

Second, it could, perhaps, be argued that, despite these limits to knowledge sharing and knowledge integration on projects, the constant dismantling of teams and redistribution of staff with experience of new ways of working might alternatively create a network of weak ties within *and* between organisations that still enables the sharing and diffusion of knowledge and new practice. According to Granovetter (1973), weak ties – which are associated with much less intimacy and emotional intensity than the close social bonding associated with strong ties – are nevertheless vitally important in providing access to wider social networks and to new and diverse sources of knowledge and expertise. However, it has also been shown that, although such weak ties may enable the transmission of explicit knowledge, they are less effective than strong ties in enabling the sharing of tacit knowledge (Hansen, 1999; Hansen *et al.*, 1999). Yet, it is precisely such tacit knowledge that is arguably crucial to more complex, longer term organisational innovation and learning. Consequently, the break up and reconfiguration of teams does little to help the sharing of knowledge through supply chains or supply networks, as it continually threatens the close bonds on which the exchange of tacit knowledge depends. As the same Property Development Director who was introduced before put it:

'Partnering is relatively new in the construction industry and it requires people to go through . . . a paradigm shift. The more in their organisation can go through that paradigm shift, the less reliant we have to be on a small core group of people. That's in the general attitude of mind. In terms of specifics, you do need a core team that you can continually use. Doesn't have to be all of them, but you need a core group. Otherwise, you do risk losing learning curve . . . There needs to be some memory.'

Project team building is, of course, important in promoting strong, local connections (i.e. strong ties) within the team that are centred upon specific project activities and objectives. Moreover, such connections often revolve around complementary or 'non-redundant' types of knowledge (e.g. inter-professional, inter-organisational), which can potentially be important in the generation of new knowledge or innovative practice (Hansen *et al.*, 1999) – assuming, that is, that the extent of interaction goes further than that associated with 'knowledge collectivities' (Lindkvist, 2005). However, such team building can also promote a degree of cohesiveness that is so strong that it effectively isolates the team from its wider organisational setting, thus further attenuating the links between project performance and wider organisational performance (Bresnen *et al.*, 2004). In other words, it is quite possible that strong project subcultures become so loosely coupled from the wider organisation that they militate further against broader (inter-)organisational innovation and learning (Dubois and Gadde, 2002; O'Dell and Grayson, 1998).

Third, research on the diffusion of knowledge and learning has emphasised the situated nature of knowledge and the difficulties in capturing learning from one context and applying it to another (Cook and Brown, 1999). The widespread diffusion of innovations in practice (across projects, organisations and even sectors) depends on their abstraction from context and their re-embedding in potentially quite different circumstances. This creates the need to translate ideas in order to apply them to different situations – often with the result that these ideas become distorted. Consequently, management ideas and practices have some 'interpretative flexibility' (Bijker *et al.*, 1987) that allows wider diffusion, but which means that the process of implementation itself can have significant implications for how such ideas and practices are made sense of and enacted (cf. Weick, 1995). Indeed, this could be said of partnering and helps explain its diverse manifestations in practice (Nystrom, 2005).

In the case of projects, the problem of translation is enormously magnified to the extent that 'each project is different' and it can become extremely difficult to diffuse knowledge and learning obtained from one project to the next, as it depends crucially upon project task and organisational circumstances. Moreover, recent research has demonstrated how the diffusion of new (management) knowledge across projects is as much, if not more, dependent upon the social learning processes involved (Newell *et al.*, 2003), so that 'reinventing the wheel' becomes an almost inevitable part of the process of embedding new ways of working. In such situations, it becomes even more difficult perhaps to be optimistic about broader innovation and learning happening within supply chains, as the emphasis is always likely to be on the recontextualisation of knowledge in local project circumstances. Recent research by the author, for instance, has demonstrated how local management practices have a powerful influence on the translation and acceptance of new management initiatives in the type of decentralised systems of working found in project-based organisational settings like construction (Bresnen *et al.*, 2005).

Indeed, when it comes to embedding knowledge and learning in organisational routines, in a project setting there is a potential mismatch between the systems and routines available to support longer term learning and those dedicated to the pursuit of shorter term, more specific project objectives (Dubois and Gadde, 2002). Studies of post-project reviews, for example, demonstrate how routine project management processes can fail to encourage or, at worst, inhibit, cross-project learning (Newell *et al.*, 2006). It may be that the use of deadlines, milestones and other time based controls can provide the 'coupling' that is needed to support wider inter-organisational communication and learning (Lindkvist *et al.*, 1998). However, such a synchronisation of organisational systems is not only difficult to achieve in a project-based setting (e.g. Sapsed and Salter, 2004), but may also be much more difficult to achieve when, as already noted, there continue to be significant difficulties in integrating systems across the supply chain (Briscoe and Dainty, 2005). More generally, such problems dovetail with other findings specific to the construction sector that have tended to show how project basing militates against organisational learning and the diffusion of new management ideas through its effects on limiting organisations' 'absorptive capacity' (Gann, 2001; Gann and Salter, 2000; Winch, 1998).

4.6   Conclusion

This chapter has looked at innovation, knowledge diffusion/sharing and organisational learning within construction supply chains, taking the analysis of the effects of conditions within the sector further than existing research tends to, by adopting a practice-based perspective that highlights the situated and socialised nature of learning processes and that explores the ways in which the particular circumstances and constraints of project-based organisation militate against the exploration (and even exploitation) of knowledge and learning within construction supply chains.

In doing so, the latter part of the discussion has highlighted the ways in which the conditions of project working and project-based organisation not only inhibit attempts to innovate and learn from establishing close supply chain relationships ('cooperating to learn'), but also how they shape and influence the diffusion and implementation of new management ideas and practices – including those concerned with SCM and partnering (Bresnen and Marshall, 2002; Bresnen *et al.*, 2005) – thus affecting organisations' abilities to put such initiatives in place and develop appropriate internal capabilities ('learning to cooperate'). The implication of this analysis is that the problems of project-based organisation and the highly situated nature of innovation, knowledge sharing, knowledge integration and organisational learning processes tend to mutually reinforce one another – in ways that reflect and reinforce the driving influence of project objectives and the desire to exploit knowledge and learning to improve immediate project performance. In other words, innovation and learning tend to default to the local and particular, with all that implies for attempts to develop new approaches and thinking within the sector more widely.

Given that existing work suggests not only that there are problems with innovation and learning within the sector, but also that SCM is, as yet, underdeveloped as an approach, it would seem that there is much to do before the full benefits of SCM within the sector for knowledge, innovation and learning can be achieved. Moreover, there is the need too for firms to deal with the dilemma that lies at the heart of their attempt to develop longer term supply chain relationships and which relates to this focus on project objectives and to the imperatives of project-based working. The earlier discussion suggested that there is no reason why the circumstances of the industry should determine action and that change is possible. However, if organisations are to develop a more explorative as well as exploitative approach to knowledge and learning, then mechanisms and approaches need to be developed that enable supply partners to partially distance themselves from the immediate demands of projects and the mindset this creates in order to gain the benefits of more explorative learning.

References

Akintoye, A., McIntosh, G. and Fitzgerald, E. (2000) A survey of supply chain collaboration and management in the UK construction industry. *European Journal of Purchasing and Supply Management*, 6(1), 159–168.

Barlow, J., Cohen M., Jashapara, A. and Simpson, Y. (1997) *Towards Positive Partnering*. Bristol: The Policy Press.

Bennett, J. and Jayes, S. (1995) *Trusting the Team: the Best Practice Guide to Partnering in Construction*. Reading: Reading Construction Forum.

Bessant, J., Kaplinsky, R. and Lamming, R. (2003) Putting supply chain learning into practice. *International Journal of Operations and Production Management*, 23(2), 167–184.

Bijker, W.E., Hughes, T. and Pinch, T.J. (eds) (1987) *The Social Construction of Technological Systems*. London: MIT Press.

Bresnen, M. (1996) 'An organisational perspective on changing buyer-supplier relations: a critical review of the evidence'. *Organisation*, 3(1), 121–146.

Bresnen, M. (2006) Conflicting and conflated discourses? Project management, organisational change and learning. In D. Hodgson and S. Cicmil (eds), *Making Projects Critical*, pp. 68–89. Basingstoke: Palgrave Macmillan.

Bresnen, M. (2007) Deconstructing partnering in project-based organisation: seven pillars, seven paradoxes and seven deadly sins. *International Journal of Project Management*, 25(4), 365–374.

Bresnen, M., Edelman, L., Newell, S., Scarbrough, H. and Swan, J. (2003) Social practices and the management of knowledge in project environments. *International Journal of Project Management*, 21(3), 157–166.

Bresnen, M., Edelman, L., Newell, S., Scarbrough, H. and Swan, J. (2004) The impact of social capital on project-based learning. In M. Huysman and V. Wulf (eds), *Social Capital and Information Technology*, pp. 231–268. Cambridge, MA: MIT Press.

Bresnen, M., Goussevskaia, A. and Swan, J. (2005) Organizational routines, situated learning and processes of change in project-based organizations. *Project Management Journal*, 36(3), 27–41.

Bresnen, M. and Marshall, N. (2000) Building partnerships: case studies of client-contractor collaboration in the UK construction industry. *Construction Management and Economics*, 18(7), 819–832.

Bresnen, M. and Marshall, N. (2002) The engineering or evolution of co-operation? A tale of two partnering projects. *International Journal of Project Management*, 20(7), 497–505.

Briscoe, G. and Dainty, A.R.J. (2005) Construction supply chain integration: An elusive goal? *Supply Chain Management: An International Journal*, 10(4), 319–326.

Briscoe, G., Dainty, A.R.J. and Millett, S. (2001) Construction supply chain partnerships: Skills, knowledge and attitudinal requirements. *European Journal of Purchasing & Supply Management*, 7(2), 243–255.

Briscoe, G., Dainty, A.R.J., Millett, S. and Neale, R. (2004) Client led strategies for construction supply chain management. *Construction Management and Economics*, 22(2), 193–201.

Brown, J.S. and Duguid, P. (2001) Knowledge and organization: A social practice perspective. *Organization Science*, 12, 198–213.

Childerhouse, P., Lewis, L., Naim, M. and Towell, D.R. (2003) Re-engineering a construction supply chain: A material flow approach. *Supply Chain Management*, 8(4), 395–406.

Christopher, M. (1992) *Logistics and SupplyCchain Management: Strategies for*
Cook, S.D.N. and Brown, J.S. (1999) Bridging epistemologies: the generative dance between organisational knowledge and organisational knowing. *Organization Science*, 10(4), 381–400.

Cox, A. (1999) Power, value and supply chain management. *Supply Chain Management*, 4(4), 167–175.

Cox, A. and Ireland, P. (2002) Managing construction supply chains: The common sense approach. *Engineering, Construction and Architectural Management*, 9(5–6), 409–418.

Cox, A., Ireland, P., Lonsdale, C., Sanderson, J. and Watson, G. (2001) *Supply Chains,Mmarkets and Power: Mapping Buyer and Supplier Power Regimes*. London: Routledge.

Cox, A. and Townsend, M. (1998) *Strategic Procurement in Construction: Towards Better Practice in the Management of Construction Supply Chains*. London: Thomas Telford.

Dainty, A.R.J., Briscoe, G.H. and Millett, S. (2001) Subcontractor perspectives on supply chain management. *Construction Management and Economics*, 19, 841–848.

DeFilippi, R. and Arthur, M. (1998) Paradox in project-based enterprises: the case of filmmaking. *California Management Review*, 40(2), 125–140.

Dubois, A. and Gadde, L.E. (2000) Supply strategy and network effects – purchasing behaviour in the construction industry. *European Journal of Purchasing & Supply Management*, 6(2), 207–215.

Dubois, A. and Gadde, L-E. (2002) The construction industry as a loosely coupled system: implications for productivity and innovation. *Construction Management and Economics*, 20(7), 621–631.

Dwyer, F.R., Shurr, P.H. and Oh, S. (1987) Developing buyer-seller relationships. *Journal of Marketing*, 51, April, 11–27.

Dyer, J. and Nobeoka, K. (2000) Creating and managing a high performance knowledge sharing network: The Toyota case. *Strategic Management Journal*, 21, 345–367.

Egan, J. (1998) *Rethinking construction*. DETR, London.

Elg, U. and Johannson, U. (1997) Decision-making in inter-firm networks: antecedents, mechanisms and forms. *Organisation Studies*, 16(2), 183–214.

Fawcett, S.E. and Magnan, G.M. (2002) The rhetoric and reality of supply chain integration. *International Journal of Physical Distribution & Logistics Management*, 32(5), 339–361.

Gann, D.M. (2001) Putting academic ideas into practice: technological progress and the absorptive capacity of construction organisations. *Construction Management and Economics*, 19(3), 321–330.

Gann, D.M. and Salter, A. (2000) Innovation in project-based, service-enhanced firms: the construction of complex products and systems. *Research Policy*, 29, 955–972.

Granovetter, M.S. (1973) The strength of weak ties. *American Journal of Sociology*, 78(6), 1360–1380.

Green, S.D., Fernie, S. and Weller, S. (2005) Making sense of supply chain management: A comparative study of aerospace and construction. *Construction Management and Economics*, 23, 579–593.

Hakansson, H. and Snehota, I. (1995) *Developing Relationships in Business Networks*. London: International Thompson Business Press.

Hansen, M.T. (1999) The search transfer problem: the role of weak ties in sharing knowledge across organizational sub-units. *Administrative Science Quarterly*, 44, 82–111.

Hansen, M.T. (2002) Knowledge networks: explaining effective knowledge sharing in multiunit companies. *Organization Science*, 13(3), 232–248.

Hansen, M.T., Nohria, N. and Tierney, T. (1999) What's your strategy for managing knowledge? *Harvard Business Review*, 77, 106–117.

Hardy, C. and Phillips, N. (1998) Strategies of engagement: Lessons from the critical examination of collaboration and conflict in an inter-organisational domain. *Organization Science*, 9(2), 217–230.

Hardy, C., Phillips, N. and Lawrence, T.B. (2003) Resources, knowledge and influence: the organizational effects of inter-organizational collaboration. *Journal of Management Studies*, 40(2), 321–347.

Harland, C. (1996) Supply chain management. *British Journal of Management*, 7, S63–S80.

Harland, C.M., Lamming, R.C. and Cousins, P.D. (1999) Developing the concept of supply strategy. *International Journal of Operations and Production Management*, 19, 650–673.

Harland, C.M., Lamming, R.C., Zheng, J. and Johnsen, T.E. (2001) A taxonomy of supply networks. *Journal of Supply Chain Management*, 37(4), 20–27.

Harland, C., Zheng, J., Johnsen, T. and Lamming, R. (2004) A conceptual model for researching the creation and operation of supply networks. *British Journal of Management*, 15, 1–21.

Holmqvist, M. (2003) Intra- and interorganisational learning processes: An empirical comparison. *Scandinavian Journal of Management*, 19, 443–466.

Holti, R., Nicolini, D. and Smalley, M. (2000) *The Handbook of Supply Chain Management: The Essentials*. London: CIRIA.

Holti, R. and Standing, H. (1996) *Partnering as Inter-Related Technical and Organisational Change*. London: Tavistock.

Holti, R. and Whittle, S. (1998) *Guide to Developing Effective Learning Networks in Construction*. London: CIRIA/Tavistock.

Knauseder, I., Josephson, P.E. and Styhre, (2007) A. Learning approaches for housing, services and infrastructure project organisations. *Construction Management and Economics*, 25(8), 857–867.

Kotabe, M., Martin, X. and Domoto, H. (2002) Gaining from vertical partnerships: Knowledge transfer, relationship duration and supplier performance improvement in the US and Japanese automotive industries. *Strategic Management Journal*, 24(4), 293–316.

Lamming, R.C. (1993) *Beyond Partnership: Strategies for Innovation and Lean Supply*. Hemel Hempstead: Prentice Hall.

Lamming, R.C. (1996) Squaring lean supply with supply chain management. *International Journal of Operations and Production Management*, 10(2), 183–196.

Latham, M. (1994) *Constructing the Team*. HMSO, London.

Lave, J. and Wenger, E. (1991) *Situated Learning: Legitimate Peripheral Participation*. Cambridge: Cambridge University Press.

Leonard-Barton, D. (1995) *Well-Springs of Knowledge: Building and Sustaining the Sources of Innovation*. Boston: Harvard Business School Press.

Levinthal, D.A. and March, J.G. (1993) The myopia of learning. *Strategic Management Journal*, 14, 95–112.

Lindkvist, L. (2005) Knowledge communities and knowledge collectivities: A typology of knowledge work in groups. *Journal of Management Studies*, 42(6), 1190–1210.

Lindkvist, L., Soderlund, J. and Tell, F. (1998) 'Managing product development projects: on the significance of fountains and deadlines', *Organisation Studies*, 19(6), 931–951.

McIvor, R. and McHugh, M. (2000) Collaborative buyer supplier relations: Implications for organisation change management. *Strategic Change*, 9(4), 221–236.

Naim, M. and Barlow, J. (2003) An innovative supply chain strategy for customized housing. *Construction Management and Economics*, 21, 593–602.

Newell, S., Bresnen, M., Edelman, L., Scarbrough, H. and Swan, J. (2006) Sharing knowledge across projects: limits to ICT-led project review practices. *Management Learning*, 37(2), 167–185.

Newell, S., Edelman, L., Scarbrough, H., Swan, J. and Bresnen, M. (2003) 'Best practice' development and transfer in the NHS: the importance of process as well as product knowledge. *Health Services Management Research*, 16, 1–12.

Newell, S., Robertson, M., Scarbrough, H. and Swan, J. (2002) *Managing Knowledge Work*. Basingstoke: Palgrave.

Nicolini, D., Holti, R. and Smalley, M. (2001) Integrating project activities: The theory and practice of managing supply chains through clusters. *Construction Management and Economics*, 19, 37–47.

Nystrom, J. (2005) The definition of partnering as a Wittgenstein family-resemblance concept. *Construction Management and Economics*, 23(5), 473–481.

O'Dell, C. and Grayson, J. (1998) If only we knew what we know: identification and transfer of international best practices. *California Management Review*, 40(3), 154–174.

Ofori, G. (2000) Greening the construction supply chain in Singapore. *European Journal of Purchasing and Supply Management*, 6(3–4), 195–206.

Phua, F.T.T. (2006) When is construction partnering likely to happen? An empirical examination of the role of institutional norms. *Construction Management and Economics*, 24(6), 615–624.

Powell, W.W., Koput, K.W. and Smith-Doerr, L. (1996) Interorganisational collaboration and the locus of innovation: networks of learning in biotechnology. *Administrative Science Quarterly*, 41, 116–145.

Proverbs, D.G. and Holt, G.D. (2000) Reducing construction costs: European best practice supply chain implications. *European Journal of Purchasing & Supply Management*, 6(3–4), 149–158.

Rothwell, R. (1994) Towards the fifth generation innovation process. *International Marketing Review*, 11, 7–31.

Saad, M., Jones, M. and James, P. (2002) A view of progress towards the adoption of supply chain management (SCM) relationships in construction. *European Journal of Purchasing & Supply Management*, 8, 173–183.

Sako, M. (1992) *Prices, Quality and Trust: Inter-Firm Relations in Britain and Japan*. Cambridge: Cambridge University Press.

Sapsed, J. and Salter, A. (2004) Postcards from the edge: Local communities, global programmes and boundary objects. *Organisation Studies*, 25(9), 1515–1534.

Styhre, A., Josephson, P.E. and Knauseder, I. (2004) Learning capabilities in organisational networks: Case studies of six construction projects. *Construction Management and Economics*, 22(9), 957–966.

Swan, J., Goussevskaia, A., Newell, S., Robertson, M., Bresnen, M. and Obembe, A. (2007) Modes of organizing biomedical innovation in the UK and US and the role of integrative and relational capabilities. *Research Policy*, 36(4), 529–547.

Sydow, J., Lindkvist, L. and DeFillippi, R. (2004) Project-based organisations, embeddedness and repositories of knowledge. *Organization Studies*, 25(9), 1475–1489.

Thorpe, A., Dainty, A.R.J. and Hatfield, H. (2003) The realities of being preferred: Specialist subcontractor perspectives on restricted tender list membership. *Journal of Construction Procurement*, 9(1), 47–55.

Vrijhoef, R. and Koskela, L. (2000) The four roles of supply chain management in construction. *European Journal of Purchasing and Supply Chain Management*, 6, 169–178.

Weick, K.E. (1995) *Sensemaking in Organisations*. London: Sage.

Wenger, E. (2000) Communities of practice and social learning systems. *Organization*, 7(2), 225–246.

Winch, G.M. (1998) Zephyrs of creative destruction: understanding the management of innovation in construction. *Building Research and Information*, 26(5), 268–279.

Wood, G.D. and Ellis, R. C. T. (2005) Main contractor experiences of partnering relationships on UK construction projects. *Construction Management and Economics*, 23(3), 317–325.

Part A

5

Marketing and Pricing Strategy

Martin Skitmore and Hedley Smyth

Pricing and marketing have largely been ignored in research on supply chains and supply chain management (SCM) in construction. The *aim* of this chapter is to address this omission and show that there is scope for improvement of SCM practice. The thrust of the argument will be that effective SCM that really adds product and service value cannot be undertaken without addressing marketing and pricing strategies over the long term. Whilst short-run improvements can be made when pricing and marketing issues are considered in tactical or 'common sense' ways, creating scope for sustained improvements through managing supply chains and clusters is likely to be constrained without providing financial and marketing strategies as a fertile context; in particular, marketing supply chain policies and pricing strategy. Therefore, addressing pricing and marketing in research on SCM not only serves to place another piece in a jigsaw, but it also changes the shape of the picture. In exploring this issue, it will also be shown that some of the current understanding of SCM in construction is deterministic both amongst advocates and critics. This chapter proposes that the way in which SCM is understood and applied requires some reappraisal; this process should continue as the ideas set out in this chapter are developed in theory and practice. The *objectives* of this chapter are to explore:

- General characteristics of SCM in construction, recognising that SCM can take several forms;
- The neglected role of marketing theory in SCM, recognising marketing as the supplier's corresponding response to customer procurement in a construction agenda that has largely been procurement driven;
- The role of pricing strategy in relation to economic and marketing considerations within SCM.

Once definitional issues have been addressed in the section below, the aim and objectives will be explored under three principle themes:

- Increased *collaboration* between parties;
- Increased *added value* for customers in the chain;
- Increased *profitability* and/or *repeat business* for suppliers in the chain.

5.1   Definitions and Difference

It is difficult to establish precise or agreed definitions in SCM, marketing, and to an extent pricing. SCM means different things to different people, with practices reflecting the diversity of meaning in construction (e.g. Green, 2006). The lack of definition is partly due to procurement being a tactical issue that has been elevated to strategic board-level decision making through practices such as SCM, partnering, and lean and agile production. Definitional differences are acceptable where they provide sources of service differentiation and hence competitive advantage in practice. It is suggested, notwithstanding the above, that Christopher's 1997 definition, devised at a point in time when SCM began to evolve from logistics theory, is still appropriate and relevant today. Christopher defined SCM as:

'. . . the management of upstream and downstream relationships with suppliers, distributors and customers to achieve greater construction value at less cost'. (Christopher, 1997)

This definition immediately challenges the pricing issue as a discrete economic factor that determines whether there is economic activity but does help to articulate how resources are allocated (Oxenfeldt, 1975), in this case management activity that intervenes across organisational boundaries to lever value. This in turn invokes the role of marketing which seeks to facilitate economic activity in advance of exchange across organisational boundaries. Marketing is involved in a range of activities to create product and service differentiation and competitive advantage. Yet marketing eludes precise definition, whether defined as creating and keeping a customer (Levitt, 1983), whether it is defined through more inclusive approaches (e.g. Kotler *et al.*, 1996; Kotler, 2000), or subject to open-ended definitions of national professional bodies and institutes of marketing. Marketing definitions are also subject to conceptual development. Theoretical and applied emphasis arises from the choice of paradigm, namely: (i) the marketing mix (Borden, 1964) based upon the mix of four ingredients (4Ps, comprising product, place, promotion and price) (McCarthy, 1964) and subsequent variants; or (ii) *relationship marketing* (Berry, 1983) developed around business-to-business (B2B) relationships and especially for intangible services, which conceives added product and service value coming through relationships (e.g. Grönroos, 2000).

While there may be basic agreement upon the definition of pricing in its broadest sense, there is controversy concerning economic theory on pricing in general and in construction in particular (Skitmore, 1989). The practice of pricing construction work has been around for a very long time indeed; the earliest known version was an *ex-post reimbursement* form to individual tradesmen, the oldest being the provision of tax or military service relief, with more recent practice including monetary payments for hours worked and the cost of materials (Thompson, 1968). The development of *contracting-in-gross*, an innovation introduced to speed up the amount of building work needed to satisfy the barrack building boom brought about in the UK by the Napoleonic Wars, resulted in two fundamental changes to this process. Firstly, management and control of the construction process passed to an overarching main contractor, to whom all payments were made for handing on to tradesmen and suppliers. Secondly, potential main contractors made tenders (bids) for work based on estimates of future costs. The successful contractor's tender price was then written into the construction contract signed by both client/owner and main contractor before the commencement of the construction work. As Chang and Ive (2007) have conceptualised, the form of client-contractor procurement and hence, contract, do matter in transaction cost and project management terms, which also has conceptual relevance along supply chains, having a potential impact upon the relative importance of SCM, marketing and price in the exchange process and project management. Overall, this method still predominates today and the compilation of tenders proceeds on the basis of subcontractor and supplier quotations, coupled with some subjective judgements by the 'pricer' concerning possible profit, risk and market factors – a practice that is identical for all contractors. Practice has largely evolved in response to a mixture of common sense and legal requirements, the theoretical basis of pricing having little influence.

Until recently, it had been assumed that standard economic theory alone would provide an adequate foundation for pricing practice. Degree courses covering estimating, for example, include estimating practice as atheoretical or the assumption that the separate study of micreconomics contains the necessary theory. Reliable pricing models have remained elusive in general (Hoffman et al., 2002). Practitioners have found scant guidance from theory on how to price in practice as theory seeks to explain economic forces rather than steer decision-making (Gabor, 1977). Empirical research supports this view (e.g. Monroe, 1990; Lichenstein and Burton, 1989; Lim and Olshavsky, 1988; Bettman, 1979 have explored different aspects; cf. Skitmore and Smyth, 2007). This agenda has been re-examined (e.g. Fine, 1974; Hillebrandt, 2000; Raftery, 1991), however, on the grounds that construction costs and market movements are just too complex and unpredictable to anticipate sufficiently well for the basic tenets of (deterministic) economic theory to hold. Some work (Runeson, 2000; Runeson and Raftery, 1998; Skitmore et al., 2006) claims complexity and unpredictability are themselves simplistic arguments. Whilst economic theory seems to offer a coherent explanation of pricing movements, it provides little guidance for practice in profit/utility maximisation, accurately predicting the consequences of departures from rational assumptions, the management of interdependencies, imperfect information and social capital (Skitmore et al., 2006).

Marketing theory has been identified as a possible means of filling this void (Skitmore et al., 2006) and recent work has begun to explore its potential for construction pricing practice (Skitmore and Smyth, 2007).

5.2   Collaboration

SCM in manufacturing represents the intervention or imposition of one firm in the activities of another firm (Womack et al., 1990). For example, a vehicle manufacturer can intervene into the management of a supplier of headlights. Such intervention can help the supplier to respond, going beyond simply producing headlights for a predetermined model. Intervention can lead to opportunities for additional cost reductions, added product value and innovation. Adding value might involve redesigning the headlight production process and may act back to change the vehicle design in order for the supplier to improve the headlight product. This is collaborative activity. Whilst it can be costly, the benefits continue for every headlight produced for the same vehicle model. The cost per unit produced is therefore very small. The added value will accrue to the vehicle manufacturer, and increased market share and profitability will accrue to the headlight supplier in some portion, usually influenced by relative market power.

In this idealised example, the supplier responded to the procurer with a marketing-cum-technical response. In terms of the marketing mix, this would be categorised as *added value* to the product as one of the 4Ps. In terms of relationship marketing there would probably be closer contact and understanding between the parties, and value is created, mediated and realised through those relationships. Part of the marketing equation is price, whether from a marketing mix or relationship marketing perspective. Where conditions in the market place change, price needs to be re-evaluated (Gates 1967; Runeson and Skitmore, 1999). This raises questions concerning whether profitability can increase for the supplier. If profitability increases, from which elements does it arise: increased profit margin from the cost savings, or increased margins on the whole product or merely on the added value element? Neoclassical economics assumes independent, single-product firms operating in differentiated markets for services. SCM invokes interdependence and differentiated services at the level of contract and service delivery. This resonates more closely with classical economics and pricing theory, for example a Marshallian approach which recognises the importance of multi-product firms (Earl, 1995; cf. Campbell, 1995) and therefore where differentiating the product/service can be used by suppliers to create price differentials and market response acts back to stimulate price stratification, which can theoretically apply to managed supply chains in construction (cf. Skitmore and Smyth, 2007).

If profitability does not increase on the product in practice, are repeat business opportunities and increased market share sufficient incentives? Powerful customers may simply drive margins down, which suppliers accept as a survival strategy or perhaps take as an opportunity to pass the cost reductions along the chain so that their own margins are protected. Where market power or leverage has an impact upon profitability, repeat business becomes a prime motive to apply relationship marketing to SCM. The benefits from repeat business include:

- Increases in market share;
- Repeat customers that are timely payers, permitting an increase in return on capital employed (ROCE) by accelerating the rate at which money is circulated, hence reducing the demand for capital;
- It is necessary in practice for the benefits in i–ii above to outweigh the costs of investing in and maintaining SCM:

○ short term costs being lower with a short terms payback when applying transaction-based procurement through a combination of relational contracting and the marketing mix
○ short terms costs being higher with no short term profit increase
yet long term costs being lower and profits higher when applying
organisational behaviour-based approach through relationship
marketing
○ a combination of (a) and (c) when transitioning from relational contracting to relationship marketing/management.

This type of activity has been conducted by many companies, including
most major vehicle producers. Chrysler adopted long-term contracts for its
suppliers in the early 1990s, reducing the number of suppliers and building
relationships with them to provide the basis for SCM (Dyer, 1996). The
marketing response of suppliers was stimulated by the customer, Chrysler,
clearly indicating its preparedness to invest in closer relationships. This
created the climate for collaboration.

SCM works differently in construction. SCM has been introduced to a
large extent by client-driven agendas that have sought continuous improvement, hence added value more closely aligned with experience received in
other sectors. Contractors have been cautious in response in order to evaluate whether there is a payback in profit and repeat business (cf. Eriksson
and Pesämma, 2007). Initial adoption by many contractors was confined to
the project level and to the next tier in the supply chain (Olayinka and
Smyth, 2007), the result being that many contractors (see also Chapter
Nine) and subcontractors have yet to develop SCM with corporate investment and on-going support to the project from the main office (Smyth,
2005; 2006) for the following detailed reasons:

• Some construction clients have shown preparedness to invest in relationships, although not all. Even though contracts may be long, this may
seem short term to clients for one-off projects. It is clients with a programme of investment that have the incentive to invest in relationships
to set the conditions for collaboration. Where this occurs, scope is
created for product and service development.
• The value that contractors, subcontractors and suppliers add to the
'product' on one project does not necessarily translate into applicability
and benefits for subsequent projects even for the same client. Therefore
the costs incurred by suppliers in making the improvements are assigned
against one project budget (or occasionally to a central budget) rather
than being distributed over many 'project units' in the way manufacturing achieves.
• The value that suppliers add to services could be replicated over many
projects and indeed this occurs for materials and component suppliers
in the chain. Although this is theoretically possible for both contractors
and subcontractors (Pryke and Smyth, 2006) and is proven in practice
on a limited scale (Smyth, 2000), it tends not to occur generally.
For many construction firms, survival is the main goal (Skitmore and
Smyth, 2007) and hence their emphasis is on trying to keep overheads,

investment (Smyth, 1985) and transaction costs to a minimum (Gruneberg and Ive, 2000). The consequence is a lack of support generally and a lack of investment in systems and procedures between head offices and projects to (a) develop standard services of consistency to replace those based upon a *personality culture* (often of blame) (cf. Smyth 2000; 2004; Pryke and Smyth, 2006; Wilkinson, 2006); and (b) to invest in services that differentiate products, for example through organisational learning and knowledge management, emotional intelligence, just-in-time production and, indeed, SCM (e.g. Smyth, 2004). This links back to the general point about investment and costs; contractors and subcontractors tend to manage their firms on a project-by-project basis, rather than as programmes that could benefit from central control and support (Smyth and Pryke, 2008).

Therefore, applying SCM to construction requires careful 'translation' in order to emulate the benefits of SCM's application in manufacturing. Translation means changing practices for construction in order to replicate the essence of SCM, which has not occurred in practice. Clients had been used to receiving added product value as a matter of course from other sectors and were tired of many contractors not even meeting the basic criteria (value added) of time, cost quality/scope. They wished the practices of other industries to be introduced in construction (e.g. Egan, 1998). The clients had the motivation (Ive, 1995; Gann, 2000) and market power (Cox and Ireland, 2006) to drive this agenda into the market, particularly for large projects and client programmes, potentially yielding repeat business opportunities for contractors as the incentive for collaboration. SCM has taken different forms in construction without translating practices to create the fundamentally key elements of SCM. The tendering system ensures that clients have leverage over contractors and, to a lesser extent, contractors over subcontractors. Client leverage exerts price pressure upon contractors and subcontractors, thus inducing limited scope to increase profit margins and to invest in SCM. Increased business therefore becomes the prime motive for contractors and subcontractors to adopt SCM. Increased business can come through repeat business or referral markets. Referred business tends not to reduce overall sales costs and is more difficult to directly attribute to SCM practices. Thus many contractors and subcontractors have directly sought repeat business, which is more in line with relationship marketing concepts. The benefits from referral and repeat business (RRB) in construction include the following:

- Covering overheads on RRB contracts permits higher margins to be charged on other projects – primarily a short-run option for both referral and repeat business;
- Resultant higher turnover increases market share for both referral and repeat business;
- Clients who are timely payers (under RRB) help accelerate the circulation of (sub)contractor working capital (capital employed on projects with, for example, a 3% margin can earn 21% in a year if the same capital is circulated seven times in the year), and investing working capital whilst

it is not being employed. This can be a significant source of profit for contractors and some subcontractors, with most relevance for repeat business clients;

- Potential opportunities to increase negotiated work, especially with repeat business clients;
- The benefits accruing from i–iv above might feasibly outweigh the costs of investing in and managing supply chains alongside the implementation of relationship marketing.

5.3   Added Value

Main contractors are predominantly operating 'buy' policies in transaction cost terms – subcontracting. This reduces their capacity to add product value directly, relying upon suppliers and subcontractors to do this. However, main contractors and the entire supply chain are, theoretically, able to add service value for the client and end-user.

A review of continuous improvement practice in construction generates a range of interesting and varied scenarios. For example, the demonstration projects implemented under the Egan and post-Egan agendas in the UK show that 74% of 119 demonstration projects for which there are records (of a total of 126 demonstration projects) have sought procurement-related improvement: early involvement, integrated teams, procurement, SCM, lean construction, partnering. Yet only five projects have adopted SCM as their primary area of improvement (Olayinka and Smyth, 2007) (see Table 5.1).

Three of the SCM projects were part of client programmes, two cases providing limited evidence to suggest that contractors were treating *projects*

Table 5.1   Demonstration projects with SCM as the primary improvement

Case	Project features	(Sub)Contractor benefits
1. Client Programme	11% reduction in capital cost	Increased client satisfaction
	25% reduction in construction time	
	Approaching 100% predictability	Lower contractor transaction costs
	Approaching zero defects at handover	Lower remedial expenditure
	Zero reportable accidents	Increased effectiveness and social improvements
2. Client Programme	Up to 30% reduction in construction time	Increased contractor ROCE
	Supply chain efficiency	Lower contractor and subcontractor transaction costs, subcontractor repeat business
	Customer focus	Client design audit and project sign-off
3. Client	Just-in-time delivery and installation	Lower client costs

as *programmes*. However, this did not appear to constitute a broader or more comprehensive programme management strategy by contractors of their projects, within which SCM and relationship management were located. The first case was generated by transferring knowledge from a previous large complex project to a current client programme. This was the only case where profit margins were discussed. All contractors benefited from repeat business. There were contractual (loss and expense) claims in one case involving the second tier, but this is tentative as no evidence that SCM practices went further than client, contractor and first tier supplier was provided.

However, SCM has performed a significant, albeit frequently secondary role within other procurement-related demonstration projects. Incidents of improvement reported below are as follows (including some double counting as several projects reported a number of procurement-related initiatives):

- Early involvement – 6 initiatives with SCM;
- Integrated teams – 12 initiatives with SCM;
- Procurement – 10 initiatives with SCM;
- Lean construction – 2 initiatives with SCM;
- Partnering – 18 initiatives with SCM, often reported as 'partnering the supply chain' following Egan (1998).

In many of these cases SCM did not penetrate beyond the client, design team and contractor (cf. Greenwood, 2001); in some cases first tier subcontractors were involved; and in only two cases was the involvement of second tier subcontractors stated. It is possible that confidentiality and competitive advantage might have constrained some reporting. However, if improvement is genuinely continuous, these players should be on to the next improvement and thus 'ahead of the game'; and as such they would be incentivised to divulge in order to benefit from reputation and profile for referral business. It would seem that clients and contractors have made less use of SCM than anecdotal evidence or claims by academe and industry might suggest, and certainly the industry as a sector has made little inroads towards the widespread adoption of SCM.

It appears that SCM seldom goes beyond the first tier of subcontractors. However, what pricing strategy could contractors adopt in order to facilitate SCM? A mechanistic analysis suggests:

- If service value can be added at no extra cost by greater contractor efficiency or lower subcontractor/supplier prices (Matthews *et al.*, 2000) then profit margins and volume of work can be maintained – pursued either through collaboration to create efficiencies amongst lower tier subcontractors, or through 'bullying' subcontractors/suppliers. This perpetuates adversarial relations in the supply chain in order to, or perhaps because of the need to, squeeze prices, and thus adds weight to the criticism that so much of the reform in construction is merely adoption of the latest management 'fashion' (e.g. Green, 2006).

- If the added value increases main contractor costs, the contractor margins will have to be reduced or passed on to maintain the same profit or volume (unless the same margin or a premium can be charged on the added value – see below). From the subcontractor/supplier's point of view at the end of the chain, there is no opportunity to pass on any extra costs other than up the supply chain (and, depending upon the leverage potential) to the main contractor, hence potentially squeezing margins up the chain. In addition, subcontractors/suppliers will naturally tend to avoid such contractors in favour of those not aiming to provide added service value to the client, unless future prospects are affected.

If *all* main contractors are compelled by market forces to provide such added service value, the logical outcome is that margins and volume will be maintained and any extra costs involved will be passed onto the client/owner. However, it is clear that this point is a long way off and market power continues to affect the scope for implementing pricing and marketing strategies along the chain.

A more realistic and less mechanistic view is that contractors and subcontractors experience a range of different behaviours, some in accordance with the experiences above and some not. Some contractors, suppliers and subcontractors will price the added value at the same rate as the non-added value (value added, which is the value representing the minimum required to meet the specification and be fit for purpose), and will potentially pursue opportunistic behaviour within any uncertainties in a project or client programme to secure a premium price as it works through in the final accounts. This is a means of 'staying ahead of the game' by entering this new form of value-added 'market' – a proactive survival strategy to defend market share as a minimum or grow share in line with competitors and manage risk exposure. Mason (2007) found that contractors directly involved with 'partnering the supply chain' were frequently requiring 4–5 subcontractors to tender from an approved list, sometimes with subcontractors only being aware in the late stages of tendering of a partnering or SCM requirement for the project. The same study found that relationships had not really improved in recent years (cf. Dainty *et al.*, 2001). Contractor/subcontractor commitment is important to give assurance to others; however, personnel are often subject to change, thus threatening continuity of service (Mason 2007; Smyth and Fitch, 2007). Combined, this confirms that pricing and marketing predominantly follow the transaction marketing mix, hence the absence of a more proactive marketing approach offered by relationship marketing. It would also suggest that relational contracting is having limited or minimal impact in many chains (cf. Eriksson and Pesämma, 2007).

On the other hand, Mason (2007) also found that there have been instances of continuous improvement in supply chains. Indeed, some specialist contractors were partnering with their own subcontractors and suppliers, sometimes even in the absence of upstream partnering. This provides better opportunities for referral and repeat business work. In addition, Potts (2007) found in his study of Terminal 5 at London Heathrow Airport that savings were shared with supply partners, thus stimulating collaboration and innovation (see also Chapter Eight). This approach had, and indeed has, the potential to improve profitability on an open-book or accounting basis by giving a greater measure of transparency and perceived fairness, thus mitigating some of the problems of opportunism based upon information asymmetry (e.g. Winch, 2002). Incentives can be divided as follows: a third to the client, a third to the contractor and a third into the project-wide pot that is distributed at the project end, with suppliers also benefiting from ring-fenced profit and an incentive scheme that rewards both early problem solving and exceptional performance.

Clearly, a range of experiences and practices exist. However, the dominant approach to SCM benefits the client, the main contractor and the first tier subcontractors. This operates in a context where repeat/referral business is the main contractor incentive, and where marketing and pricing remain transaction-based.

Much of the literature conceptualising SCM assumes coherence in the chain; understanding and practices at the client–contractor interface are experienced equitably along the whole chain in price terms. Yet, such assumptions are unrealistic for construction, as contractors are reluctant to invest in new practices for the reasons cited. Therefore, change is going to be cautious and incremental. Contractors learnt the SCM rhetoric and told clients of a willingness to cooperate. In practice, they began to react as circumstances arose before and during contracts. Prior investment tended to be negligible or absent. However, this, coupled with the predominance of 'buy' strategies of subcontracting amongst main contractors, has led them to simply take on board the procurement-driven agenda from clients and pass it along the chain:

> '. . . the procurement drivers from the client are received by the contractor as a procurement model, which is emulated. Supply chain management therefore becomes characterised as procurement push, which is driven along the chain, especially where suppliers experience similar overall market conditions to the main contractors.' (Smyth, 2005: 25)

Hence, a corresponding *marketing pull* is not exerted within their own organisations, with responsibility being handed to other parties. This is part of the defensive approach for survival and thus helps contractors to keep investment and transaction costs to a minimum.

Added value as a concept is frequently defined in a loose conceptual sense to incorporate cost reduction – strictly speaking an improved value for money equation rather than addition. Where cost reduction is the client requirement, the main contractors can respond by driving down prices along the chain. This fits with the procurement pull model and accords with some common experiences of SCM in practice – contractors 'bullying' subcontractors into reducing costs. Logically, this type of SCM is likely to encourage the lengthening of chains, the use of self-employed labour and the informal economy of illegal practices, where those squeezed at the end have no understanding of the drivers for continuous improvement. It has been shown that this is an outcome from the deterministic pricing analysis viewpoint. Perhaps this is 'acceptable' from the construction client perspective as they tend not to intervene along the chain more than two links – the main contractor and first tier subcontractors – and thus, have minimal understanding of the consequences of cost reductions. Whilst there may be some *lean* practices to eliminate waste (cf. Koskela, 2000), the main thrust is to reduce tender prices (cf. Green, 2006). Whilst relational contracting provides a less adversarial context, within which SCM can be located, it is still transaction-based; hence relational contracting will only serve to ameliorate or 'soften', rather than overcome the tendencies analysed (cf. Cox and Ireland, 2006). Relational contracting provides some conceptual basis for cooperation across projects in response to client programmes (Smyth and Pryke, 2008). In practice, contractors and subcontractors have tended behave cooperatively (cf. Eriksson and Pesämma, 2007), yet not bundle projects into programmes for particular clients or for their own corporate effectiveness. On the one hand, this is due to the reluctance to invest in a transaction-based market, even a cooperative one, and on the other hand, will, vision or perhaps understanding of the potential to organisationally manage the market through relationship marketing and relationship management.

The price reduction analysis (whether as advocates or critics) is also mechanistic. Firstly, as argued, it ignores marketing for managing and potentially changing the market, especially relationship marketing (Smyth, 2005; cf. Skitmore and Smyth, 2007). Secondly, it ignores the benefits derived from SCM as evidenced by the demonstration projects (Olayinka and Smyth, 2007) and other studies (e.g. Mason 2007; Potts, 2007), even if the benefits mainly accrue to the contractors and first tier suppliers and subcontractors. The result has been a procurement-driven agenda with, broadly, three possible market responses from contractors and subcontractors:

- *(Common response)* Ignore the agenda, especially where repeat business is perceived as insufficient incentive or where there are too few referral and repeat business opportunities, with a consequential risk of losing competitive position if others respond positively.
- *(Common response)* Learn the procurement rhetoric and pass it along the supply chain for others to make the response (Smyth, 2005). This is a particularly strong option where the contractor is not 'making' anything and is 'buying' everything through the subcontracting system in transaction cost terms (cf. Gruneberg and Ive, 2000) and wishes to maintain as low a risk profile as possible.
- Embrace the procurement drivers and meet these with one of the following:
  ○ *(Common response)* A purely cost cutting, risk management approach;
  ○ *(Reasonably common response)* A project specific or, sometimes, a client programme specific response that is underpinned by a transaction marketing mix response to match certain requirements;

○ *(Rare and potential response)* A relationship marketing response with a pricing strategy that is part of the contractor corporately running its projects as a programme with SCM.

5.4   Profitability and Repeat Business

In order to replicate the essence of SCM in manufacturing as fully as possible to accrue similar benefits in construction, there needs to be a corporate level response which effectively infuses the changes into all projects through contractor programme management. It was seen that an innovation, for example in headlight design and production, accrues with every headlight produced. However, a change at the project level does not automatically transfer to other projects without corporate investment and support. One important starting point for effective SCM is the marketing strategy that extends beyond a narrow marketing mix approach of price dominated competition. Relationship marketing adds value through relationships and tries to change and manage the market through aggregated behaviour (e.g. Grönroos, 2000; Gummesson, 2001 concerning the Nordic School; Ford *et al.*, 2003 concerning the IMP Group with its industrial, manufacturing and B2B marketing focus; Berry, 1983 concerning the more pragmatic North American tradition; Smyth, 2000; 2005 in construction; cf. Skitmore and Smyth, 2007; Smyth and Edkins, 2007; Smyth and Fitch, 2007). Aggregated behaviour is changed through managing behaviour through a systematic approach, trying to replicate at corporate level, and project level for construction, the personal relationships that can be built up between sole trader and customer. It therefore requires investment in systems and procedures, supported by a customer/client-orientated organisational culture. This provides a basis for relationship marketing between organisations.

Relationship marketing constitutes a response to add value, especially service value where the contractor is managing rather than directly producing the constructed facility – the 'buy' rather than 'make' decision. In an SCM context, this dovetails well with intervention into the management of other firms in the chain. This is one way in which relationship marketing evolves into relationship management (e.g. Grönroos, 2000; Smyth and Edkins, 2007 in construction). Relationship marketing and management can therefore be used to lever service improvement and generate greater levels of collaboration, to the extent that ways of working can be changed in similar ways to changes in product manufacturing. This approach creates potential for developing trust (Smyth, 2008a) as it is relationship-based rather than a procurement driven transaction approach where trust has been empirically shown to be difficult to foster compared to cooperation (Eriksson and Pesämma, 2007), although it has also been empirically shown that trust is necessary to maintain long-term relationships for effective service delivery on projects (e.g. PPP/PFI projects) and programmes (Smyth and Edkins, 2007).

Relationship marketing and management are located within the relationship paradigm of the management of projects (Pryke and Smyth, 2006), recognising that value is added through relationships rather than the project management tools and techniques *per se*. Relationship marketing and management can be powerful means within this paradigm and conceptually add considerable scope for improvements in SCM. At this point it is important to explicitly distinguish between *relational contracting*, which can also be located within the relationship paradigm of managing projects yet has its conceptual roots within the transaction cost tradition (Macneil, 1974; cf. Williamson, 1985); and *relationship management*, which has its roots in relationship marketing. Instead of trying to reactively change behaviour by changing market structures and governance from the top-down, relationship marketing and management try to change the market proactively from the bottom-up by changing behaviour. However, it is managed behaviour aggregated at the level of the firm which helps to reduce transaction costs through more collaborative and effective working, stimulated through investment to induce the changes.

Such investment presents a *barrier to entry* to competition, with further barriers being created according to the way relationship marketing and management are configured to differentiate service provision and match clients' requirements (see Smyth 2000; 2008b; Pryke and Smyth, 2006; Smyth and Fitch, 2007; Skitmore and Smyth, 2007). If SCM is applied with relationship marketing to move beyond both cost cutting and relational contracting, towards more substantive changes that add service value and lead to differentiated services, then pricing strategies can be reappraised.

In addressing this, it is important not to fall into the constraints of mechanistic analysis, having already recognised the economic and behavioural forces that encourage contractors and subcontractors to be cautious and induce change incrementally. Therefore a transition from current practice, largely rooted in relational contracting, towards one based on relationship management is presented. Short-term efficiency tends to provide a starting point, contractors and subcontractors potentially shifting strategy in the following ways:

- From endeavours aimed at minimising risk and reducing costs, towards those which invest in relationships to add value along the supply chain;
- From a simple enterprise-orientated motivation to secure repeat business, towards organisational learning improve effectiveness and induce long-term efficiencies, coupled with a reduction of long-term transaction-based business development costs to relieve price pressures arising from relationship investment costs;
- From reactive response to clients towards added service value in order to improve value in relation to price on a project-by-project basis towards planning and developing generic areas for adding value across all projects that are specifically tailored within individual projects;
- From implementing SCM in the next tier of the chain towards managing the total length of chains and the product/service content for each link in terms of added value;

- In summary, from relational contracting to relationship marketing and management in SCM.

Less obvious are the economic gains where SCM might offer greater collaboration (partnering and agile production) at the expense of competition, shifting:

- From subcontractor/supplier collaboration with contractors that may push up prices short term (increased investment in relationships) towards time-lagged reduction in long-run new business selling and project transaction costs;
- From low subcontractor margins towards pricing strategies to earn higher margins, hence pushing up prices especially in the long run, taking into account that the cost and relationship benefits must already have been accruing to contractors or the next tier in order not to erode their price position in the chain;
- From one-off added product/service value towards repetitive added value across projects on the one hand, and exploration of new areas for added value on the other, whereby the rate of introduction is tailored to value what clients want and are prepared to pay for. This is an understanding that can only be gained from close relationships commencing with the marketing and sales process.

Therefore, relationship marketing requires a different pricing strategy:

The change can start with raising prices in a buoyant market, the return being invested to deliver the added value to increase client satisfaction and repeat business. In a steady market, then the investment has to be made first, so that the added value is demonstrated to specific clients and in referral markets through promotion and reputation before prices can be raised. Investment therefore will initially lead to an increase in working capital and a reduction in ROCE. As repeat and referral business increase, further investment can be covered, for it costs over five times more to find a new client than keep an existing one. (Skitmore and Smyth, 2007: 628)

Relationship marketing offers scope for contractors and subcontractors to add service value. Adopting this becomes complex because traditional contracting and current transitional applications of continuous improvement exhibit a range of SCM practices. The scope of relationship marketing is greater than relational contracting, which provides a basis for transition towards relationship marketing and management (Skitmore and Smyth, 2007); this can lead to reappraising pricing strategy. There is no automatic or mechanistic transition to a relationship management approach. Indeed, this analysis has drawn upon other work to show the predominance of survival approaches in construction (e.g. Skitmore *et al.*, 2006), despite the scope. However, there are contractors that have pursued (Smyth, 2000) and are pursuing relationship marketing (Smyth and Fitch, 2006) to greater and lesser degrees of rigour and success (Smyth, 2000; Edkins and Smyth, 2007).

The scope for continuous improvement via SCM is constrained by the use of cost cutting, 'bullying', and the adoption of procurement-driven agendas that lead to marginal and short-term improvements through relational contracting. A shift towards relationship marketing is a move away from survival strategies. This move carries risk as (a) new investments are made, and (b) service improvements have to be delivered and perceived in the market in order to change pricing strategies. However, the increase in repeat business and reduced transaction costs can off-set risks. Contractors and subcontractors making the transition can pursue it cautiously and incrementally, but they need to undertake it coherently and consistently to avoid the pitfalls of piecemeal approaches to relationship marketing, which have led to self-imposed constraints that affect repeat business, pricing and frequently profitability. It is the largest contractors and subcontractors that receive the greatest demands from clients, and have the resources and potential for service differentiation.

5.5   Conclusion

Research has largely left pricing and marketing unexplored, both in terms of supply chains and SCM for enterprises at any point in the chain. This chapter has examined pricing and marketing, considering market pressures that restrict opportunities for efficient and effective SCM, replicating the essence of SCM in manufacturing. Organisational constraints to formulating and implementing pricing strategies for main contractors and subcontractors within a managed chain have also been discussed.

The analysis has shown that the predominance of transaction considerations on the one hand, and maintenance of low investment for flexibility reasons on the other, has led to a price-dominated form of the marketing mix. The marketing mix is transaction-based. Applying SCM within a risk minimisation and transaction context leads to some added value and/or cost cutting, either through 'bullying' or, at best, through transaction-based relational contracting. The limits for continuous improvement are soon reached and do not measure up to the corresponding achievements in manufacturing.

A relationship marketing approach has been explored conceptually, which offers possibilities for implementation in construction supply chains. This could create further scope for contractors and subcontractors, yielding further benefits to the client in terms of added service and product value, and generating higher referral and repeat business levels along with greater subsequent profitability. This requires a corresponding pricing strategy. However, implementation of such changes in pricing strategy has been slow in practice. Whilst the analysis has shown the conceptual possibility, and also traced a transition towards a relationship marketing and pricing strategy for SCM, it largely remains 'theory' at this stage. A non-mechanistic analysis has been adopted, on which basis it would be logical to conclude that the 'jury is out' in terms of whether clients, contractors and subcontractors will choose to drive through further change, or whether the continuous improvement agenda and arguably the concept of SCM itself becomes a mere phase or passing industry fashion.

A buoyant or stable market in demand creates fertile conditions for change. However, it has also been argued that contractors and subcontractors have insufficient motive (Ive, 1995; Gann, 2000). As long as there is plenty of work, subcontractors need not be too concerned about relationships, even if contractors are. They may become concerned in a constrained market, but do not have the resources to develop such strategies for survival reasons.

This chapter has posed a challenge to contractors and subcontractors. In conclusion, the authors believe there is probably insufficient motivation for contractors and subcontractors to change strategies regarding business generally and SCM in particular, with short-term survival considerations continuing to dominate for the foreseeable future without strong client drivers. Client drivers could take a number of forms, the following being two of the most likely:

- Clients continue to drive the continuous improvement agenda generally, and specifically determine to intervene further into the affairs of contractors and directly into the affairs of all key subcontractors, with marketing measures that could include:
  - Clients requiring identifiable, hence named and stable supply chains;
  - Contractor promises made at pre-qualification and bidding stages being considered as much part of the requirements as meeting contract clauses;
  - Codes of behavioural conduct being introduced throughout supply chains;
  - Determining and specifying maximum chain length for effective SCM based upon adding value rather than solely cutting costs by squeezing formal (legal working) and informal labour (illegal working) at the far end of the chain;
  - Specified marketing responses in terms of continuity and consistency of service within projects and across project programmes;
  - Specified marketing responses for key areas of added product and service value, including accountability as to how improvements are being collaboratively created and delivered.
- Clients create economic incentives to encourage efficient and effective SCM:
  - Clients acknowledge the need for chains to achieve reasonable levels of repeat business;
  - Clients act ethically (not opportunistically to use their market power to capture all financial benefits) to permit equitable distribution of financial benefits along supply chains (cf. Potts, 2007; Smyth, 2007; Smyth, 2008a), working with contractors and subcontractors to set 'ground rules' on pricing strategies for genuine mutual benefit (as opposed to rhetorical 'mutuality' where benefits accrue 6 : 1 in favour of clients, as set out by Bennett and Jayes, 1995);

○ Contract terms, including promises, between client and contractor are replicated throughout the supply chain, more onerous terms not being permitted;

○ Financial incentives comprise part of the required terms and promises;

○ The documentation of (sub-)contractors' (and consultants') good practice SCM track record, and the inclusion of this in the contractor and consultant selection process.

Large clients with investment programmes are those with the motives to introduce such measures, whilst professional and trade bodies have an indirect role as advocates of 'best practice' and a direct role in their production of codes of practice. If a number of clients collaborate on these principles in the same way that the continuous improvement agenda was introduced, then some of the largest contractors would benefit and demand will be concentrated in their hands. Such capital concentration is a normal economic process in other industries and could create a 'super-league' of contractors and subcontractors, creating barriers to entry. This would change a prominent part of the sector.

This chapter has presented what is possible, showing the opportunity for contractors and subcontractors to take up the challenge and respond out of a sense of vision and reward for their clients and all members in the chain, yet also recognising the reasons why this challenge may not be taken up. The chapter has, therefore, concluded with an outline of how the conditions can create a shift in SCM practices, taking into account that anything that involves spending more money, even as an investment, decreases the risk of short-term survival of the firm.

References

Bennett, J. and Jayes, S. (1995) *Trusting the Team*. Reading Construction Forum, Centre for Strategic Studies in Construction, University of Reading, Reading.

Berry, L.L., Shostack, G.L and Upah, G.D. (eds). (1983) Relationship Marketing. *Emerging Perspectives on Service Marketing*. American Marketing Association, Chicago.

Bettman, J.R. (1979) *An Information Processing Theory of Choice*, Addison-Wesley, Reading, MA.

Borden, N. (1964) The concept of the marketing mix, *Journal of Advertising Research*; June, 2–7.

Campbell, N. (1995) *An Interaction Approach to Organisation Buying Behaviour: relationship marketing for competitive advantage*. Butterworth-Heinemann, Oxford.

Chang, C-Y. and Ive, G. (2007) The principle of inconsistent trinity in the selection of procurement systems, *Construction Management and Economics*, 25, 677–690.

Christopher, M. (1997) *Marketing Logistics*. Butterworth Heinnemann, Oxford.

Cox, A. and Ireland, P. (2006) Relationship management theories and tools in project procurement *Management of Complex Projects: a Relationship Approach*, Pryke, S.D. and Smyth, H.J. (eds.), Blackwell, Oxford.

Dainty, A., Briscoe, G. and Millett, S. (2001) Subcontractor perspectives on supply chain alliances, *Construction Management and Economics*, 19, 841–848.

Dyer, J.H. (1996) How Chrysler created an American keiretsu, *Harvard Business Review*, 74(4), 42–56.

Earl, P.E. (1995) *Microeconomics for Business and Marketing*, Edward Elgar, Cheltenham.

Egan, Sir John (1998) *Rethinking Construction*, HMSO, London.

Eriksson, P.E. and Pesämma, O. (2007) Modelling procurement effects on cooperation. *Construction Management and Economics*, 25, 893–901.

Fine, B. (1974) Tendering strategy. *Building*, 25 October, 115–121.

Ford, D., Gadde, L-E., Håkansson, H. and Snehota, I. (2003) *Managing Business Relationships*, Wiley, Chichester.

Gabor, A. (1977) *Pricing: Principles and Practices*. Gower, Aldershot.

Gann, D. (2000) *Building Innovation*. Thomas Telford, London.

Gates, M. (1967) Bidding strategies and probabilities. *Journal of the Construction Division, American Society of Civil Engineers*, 97(CO2), 277–303.

Green, S.D. (2006) Discourse and fashion in supply chain management, *The Management of Projects: a Relationship Approach*, Pryke, S.D. and Smyth, H.J. (eds.), Blackwell, Oxford.

Greenwood, D. (2001) Sub-contract procurement: are relationships changing? *Construction Management and Economics*, 19, 5–7.

Grönroos, C. (2000) *Service Management and Marketing*. John Wiley and Sons, London.

Gruneberg, S.L. and Ive, G.J. (2000) *The Economics of the Modern Construction Firm*, Macmillan, Basingstoke.

Gummesson, E. (2001) *Total Relationship Marketing*, Butterworth-Heinemann, Oxford.

Hillebrandt, P.M. (2000) *Economic Theory and the Construction Industry*, Macmillan, Basingstoke.

Hoffman, K.D., Turley, L.W. and Scott, W.K. (2002) Pricing retail services, *Journal of Business Research*, 55, 1015–1023.

Ive, G. (1995) The Client and the Construction Process: the Latham Report in Context. *Responding to Latham: the views of the construction team* (ed. S.L. Gruneberg), CIOB, Ascot.

Koskela, L. (2000) *An Exploration towards a Production Theory and its Application to Construction*, Report 408, VTT, Espoo.

Kotler, P. (2000) *Marketing Management*, Prentice Hall International, Englewood Cliffs.

Kotler, P., Armstrong, G., Saunders, J. and Wong, V. (1996) *Principles of Marketing*, Prentice Hall, Hemel Hempstead.

Levitt, T. (1983) *The Marketing Imagination*, Free Press, New York.

Lichtenstein, D.R. and Burton, S. (1989) The relationship between perceived and objective price quality. *Journal of Marketing Research*, 26(4), 429–443.

Lim, J-S. and Olshavsky, R.W. (1988) Impacts of consumer's familiarity and product class on price-quality inference and product evaluations. *Quarterly Journal of Business Economics*, 27(3), 130–146.

McCarthy, E.J. (1964) *Basic marketing: a managerial approach*. Richard D Irwin Inc.

Macneil, I.R. (1974) The many futures of contracts. *Southern California Law Review*, 47(3), 691–816.

Mason, J. (2007) Specialist contractors and partnering. *Collaborative Relationships in Construction*, Smyth, H.J. and Pryke, S.D. (eds.), Blackwell, Oxford.

Matthews, J., Leah, P., Phua, F. and Rawlinson, S. (2000) Quality relationships: partnering in the construction supply chain, *International Journal of Quality and Reliability Management*, 17(4–5), 493–510.

Monroe, K.B. (1990) *Pricing-Making Profitable Decisions*. McGraw-Hill, New York.

Olayinka, R. and Smyth, H.J. (2007) Analysis of Types of Continuous Improvement: demonstration projects of the Egan and post-Egan agenda. *Proceedings of RICS Cobra 2007*, 6–7 September, Georgia Institute of Technology, Atlanta.

Oxenfeldt, A.R. (1975) *Pricing Policy*. AMACOM, New York.

Potts, K. (2007) Change in the quantity surveying profession: Heathrow Terminal 5 case study. *Collaborative Relationships in Construction*, Smyth, H.J. and Pryke, S.D. (eds.), Blackwell, Oxford.

Pryke, S.D. and Smyth, H.J. (2006) *The Management of Projects: a relationship approach*. Blackwell, Oxford.

Raftery, J. (1991) *Principles of Building Economics*, BSP Professional Books.

Runeson, G. (2000) *Building Economics*. Deakin University Press.

Runeson, G., Raftery, J. (1998) Neo-classical micro-economics as an analytical tool for construction price determination. *Journal of Construction Procurement*, 4(2), 116–131.

Runeson, G. and Skitmore, R.M. (1999) Tendering theory revisited. *Construction Management and Economics*, 17(3), 285–296.

Skitmore, R.M. (1989) *Contract Bidding in Construction: Strategic Management and Modelling*. Longman Scientific and Technical, Harlow.

Skitmore, R.M., Runeson, G., Xinling Chang. (2006) Construction price formation: full-cost pricing or neoclassical microeconomic theory? *Construction Management and Economics*, 24(7), 773–784.

Skitmore, R.M. and Smyth, H.J. (2007) Pricing construction work: a marketing viewpoint, *Construction Management and Economics*, 25(6), 619–630.

Smyth, H.J. (1985) *Property Companies and the Construction Industry in Britain*. Cambridge University Press, Cambridge.

Smyth, H.J. (2000) *Marketing and Selling Construction Services*. Blackwell, Oxford.

Smyth, H.J. (2005) Procurement push and marketing pull in supply chain management: the conceptual contribution of relationship marketing as a driver in project financial performance. *Journal of Financial Management of Property and Construction*, 10(1), 33–44.

Smyth, H.J. (2004) Competencies for improving construction performance: theories and practice for developing capacity. *The International Journal of Construction Management*. 4(1), 41–56.

Smyth, H.J. (2006) Competition, *Commercial Management of Projects: Defining the discipline*. Lowe, D. and Leiringer, R. (eds.), Blackwell, Oxford.

Smyth, H.J. (2007) The moral economy and an ethics of care perspective in relationship marketing. *Proceedings of EURAM 2007*, May 16–19, INSEAD, Paris.

Smyth, H.J. (2008a) Developing trust. *Collaborative Relationships in Construction: Developing Frameworks and Networks*. Blackwell, Oxford.

Smyth, H.J. (2008b) The credibility gap in stakeholder management: ethics and evidence of relationship management. *Construction Management and Economics*, 26(6), 633–643.

Smyth, H.J. and Edkins, A.J. (2007) Relationship management in the management of PFI/PPP projects in the UK. *International Journal of Project Management*, 25(3), 232–240.

Smyth, H.J. and Fitch, T. (2007) Relationship management: a case study of key account management in a large contractor, Paper presented at *CME25: Construction Management and Economics: past, present and future*, 16–18 July, University of Reading, Reading.

Smyth, H.J. and Pryke, S.D. (2008) *Collaborative Relationships in Construction: developing frameworks and networks*. Blackwell, Oxford.

Winch, G.M. (2002) *Managing Construction Projects*. Blackwell, Oxford.

Thompson, F.M.L. (1968) Chartered Surveyors: The Growth of a Profession, London: Routledge & Kegan Paul.

Wilkinson, S. (2006) Client handling models for continuity of service, *Management of complex projects: a relationship approach*. Pryke, S.D. and Smyth, H.J. (eds.), Blackwell, Oxford.

Williamson, O.E. (1985) *The Economic Institutions of Capitalism*. Free Press, New York.

Womack, J., Jones, D. and Roos, D. (1990) *The Machine that Changed the World*. Rawson, New York.

Part A

Part B

Application and Case Studies

This section of the book takes a very practical look at the application of supply chain management (SCM) in construction. The first chapter is an interesting contribution from Andrew Edkins which deals with risk in project supply chains and commences the discussion of the applied work contained in Part B of the book. The next two chapters in this part look at the experiences and achievements of two of the largest construction clients in the UK – Slough Estates and British Airports Authority. The chapter on Slough Estates is contributed by Bernard Rimmer, ex-Construction Director at Slough Estates and architect of many of the most important SCM initiatives introduced by Slough Estates. Keith Potts worked with British Airport Authority (BAA) senior staff to provide us with an overview of the very important contribution that BAA made to supply chain thinking in UK construction. Having looked at the application of SCM from the clients' viewpoint, we have an important contribution from Andrew King and Martin Pitt, which deals with the way in which the contractor might exploit SCM. Finally, Hedley Smyth provides us with an account of the use of franchising in construction. Franchising might apply to both contractor and specialist subcontractors and the chapter completes Part B, providing a balance to the various interests represented in the construction supply chain.

Risk Management and the Supply Chain

Andrew Edkins

6.1 Introduction

This chapter will explore the use of supply chains and their management, and the inextricably linked challenge of managing risk. This chapter provides an insight into some of the more subtle distinctions made between the rhetoric and practice of managing risk through the use of supply chains. It will consider what the difference is between risk and uncertainty – and whether such a distinction is important; and then consider how the practical challenge of forming and managing a supply chain may result in unexpected consequences.

We first need to explore the issue of risk. Risks are present all the time and everywhere. For those who derive a living from being expert in risk management, there is a specialist area of knowledge with its own lexicon of terms. For the purposes of this chapter, the specific terms to be used need careful definition. 'Risk' henceforth will be considered as comprising risk and uncertainty. The word 'uncertainty' is typically used to indicate less confidence, suggesting a class of vagueness that is inferior to those concerns that can be tagged as 'risks', where there may be more formal assessment of both impact and likelihood of the risk, risk event or hazard (Knight, 1921). In this context, the Oxford English Dictionary defines uncertainty as: '[The quality of] a business risk which cannot be measured and whose outcome cannot be predicted or insured against' (www.dictionary.oed.com, accessed March 2008). The use of the word risk to cover both terms reflects the general mood of modern literature, which is dominated by the term risk. However, in the context of the study of risk, it is important to be clear about the nature of the risk to be managed, and therefore the distinction between risk and uncertainty should be at least considered.

In the world of risk management there is great importance placed on the assessment of issues associated with terms such as 'impact' and 'likelihood'. Combining the two allows the use of a risk register where a particular risk (event) has a consequence (impact) with an assessed probability (likelihood). Such risk registers are a mainstay of project risk management and have been developed to very high levels of sophistication, depending upon the nature of the project (context) and who is involved in generating and reviewing the register.

The term 'uncertainty management' is not as familiar as 'risk management', as the implied vagueness requires more general use of common-sense approaches. The Harvard Business Review's (1999) useful summary of texts under the title *Managing Uncertainty* deal with a number of subjects, but with an overall emphasis on setting forward-looking strategy and business planning. This leads to the second important point – that uncertainty is frequently associated with benefits or opportunities and much of the discussion about the management of *risk* (intrinsically downside biased) can be applied to the management of *opportunity* – considered as the potential benefit. For example, many of our most personally cherished memories are from occasions or events that have some element of thrill or excitement. Indeed, there are those in society that we can call 'thrill-junkies', or more formally 'risk seekers', who enjoy pastimes that many others would consider as being extraordinarily dangerous or scary. In business, many investment decisions are taken for the potential pay-off or benefit(s), recognising that the investment may fail, but hoping that the proverb – *speculate to accumulate* – will prove true. Bringing the two issues of risk and benefit management together provides us with the management challenge shown below in Figure 6.1.

In considering Figure 6.1, we can imagine that a manager has a mental radar screen that is scanning for risks and uncertainties. The radar screen is an often used metaphor for very good reason. It suggests a proactive set

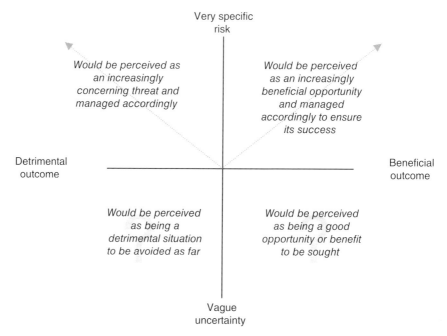

Figure 6.1   The risk/uncertainty management challenge.

of activities that involves receiving signals that need assessing, and then taking action. Whilst modern radar may have a very high degree of richness about the data it displays, the simple mental image of a screen with a sweeping hand that reveals the presence of an object or entity is critical to the metaphor, as it leaves the observer to determine what the object is and what to do about it. Some signals are very welcome, such as the identification of the destination landscape, whilst others may be direct and imminent threats, such as an advancing enemy or object on a collision course. For the purposes of this chapter, the two key requirements of the use of this managerial radar are the need for it to be resourced so that the signals received are being brought to the attention of those capable of taking action, and that there is sufficient skill and competence to interpret the information arising and implement the corresponding actions.

These points take on a centrally important role in the context of the project, especially when the project is delivered through the use of various organisations working as a supply chain (Smith, 1995). Figure 6.1 illustrates the two different types of challenge that are faced by managers and leaders of projects and the supply chain organisations (SCOs) that work on them; some issues may be perceived as *vague* concerns (those that would be considered below the horizontal axis) and others are deemed *specific* risks (above the axis) (Knight, ibid, Keynes, 1921). Since there is debate as to whether something is a risk or an uncertainty, it is often the case that risks are those that are recorded in list form and given further treatment, whilst uncertainties may remain in the background, possibly with many aware of them, but never quite tangible enough to make it into a formal record. There is significant evidence from reviews of project failures to show that such negative uncertainties (detrimental outcomes) were known by many, but not acted upon as they were not capable of being analytically defended at the time (Flyvbjerg *et al.*, 2003, Miller and Lessard, 2000).

Such was the case in the US Challenger space shuttle tragedy in 1986, when the decision to launch the space shuttle mission was taken despite the prediction of record low temperatures that were outside the parameters encountered previously. This tragic accident was reviewed by a Presidential Commission that amongst other findings reported:

'The decision to launch the Challenger was flawed. Those who made that decision were unaware of the recent history of problems concerning the O-rings and the joint and were unaware of the initial written recommendation of the contractor advising against the launch at temperatures below 53 degrees Fahrenheit and the continuing opposition of the engineers at Thiokol [Morton Thiokol were the manufacturers of the shuttle's solid booster rocket – SRB]) after the management reversed its position. They did not have a clear understanding of Rockwell's [the shuttle's prime contractor] concern that it was not safe to launch because of ice on the pad. If the decision-makers had known all of the facts, it is highly unlikely that they would have decided to launch 51-L on January 28, 1986.' (Presidential Commission, 1987)

Here, we have a series of organisations working in a supply chain, each with specific knowledge, competences and skills. This type of project coalition is common across many sectors and it would be considered normal for a UK construction project to have a series of expert organisations linked via contracts, each bringing expertise with accompanying skills and competences.

However, whilst some may be familiar supply chains, their identification as such is not universal. Supply chains are created specifically by linking those individuals and organisations together in a manner that creates value in excess of the additional costs associated with maintaining the supply chain. Economists would counter supply chains with the use of vertically integrated businesses that effectively provide the same solution, but from within the confines of one organisation (Coase, 1937). The rise of the concept of the supply chain has, arguably, been largely a reflection of the dynamism and flexibility needed to exploit rapidly changing markets or specific opportunities. The UK construction industry is now heavily dependent on the use of supply chains, supplying up to 80% of the construction project's value (Constructing Excellence, undated). Only time will tell whether supply chains will be as dominant in the future, but in the meantime we can investigate some of today's specific challenges facing supply chains and their management.

6.2   Placing the UK Construction Industry in Context

The UK economy, like any other, must adapt to both the external and internal environments. The development of construction procurement can be seen to be moving on a more evolutionary basis than a revolutionary one. Even such modern developments as the internet can be interpreted in one sense at least as initially providing nothing more than a virtual shop window to customers. Yet with modern societal and economic developments, there has come an increasing realisation that the purchase of goods and services carries with it both opportunity and responsibility. In unpacking this point, first consider the default option – doing everything yourself. For the vast majority of society in countries like the UK we are very far from being individually self-sufficient. We are now more likely to buy than make and this is the same for the corporate world. Whilst there has been industrial development for some centuries now, modern companies may increasingly be described as 'niche' or 'portfolio', and act accordingly. This is a departure from the vertically-integrated structure, where companies like the Ford Motor Corporation, particularly in the very early days of mass production, not only carried out all the manufacture and assembly operations, but also sourced most materials directly from their own extractive operations and sold one final product through their own dealerships. Such completeness of vertical integration reached its zenith in the mid part of the twentieth century.

Before the focus becomes totally construction specific, it is worth noting that the challenges of construction are not unique. Construction is one example of a project-dominated industry, but there are others such as ship-building and that portion of the creative media industry associated with film and TV production. Other industries, such as aerospace, petrochemicals, pharmaceuticals and IT can also rely heavily on projects, but also deliver products via operational or production management modes. In all cases cited above, increased fragmentation has been observed (Hillebrandt, 2000; Hillebrandt andand Cannon, 1990). Sometimes, as in the case of pharmaceuticals, this is more embryonic, with still much resource being retained in-house by the pharmaceutical giant corporations. However, the development of gene-mapping and biotechnology has heralded opportunities for new companies to explore highly specialist areas, either associated with particular clusters of biological elements, or with highly complex parts of the process of understanding and adapting such molecules and biological agents. The point here is that increased specialisation almost inevitably involves the employment of highly specialised service providers and their management.

It is significant when comparing and contrasting these project-based industries to consider how 'fixed' the projects are in terms of location. Shipbuilding typically produces the bulk of the project in a yard. These yards are fixed in location, so ships are produced in places and when sufficiently complete can then travel to continue specialist fit out in other places. Film and TV production require both studios and in some cases specific locations, so are partially dependent upon specialist space. Construction has traditionally been centred on the final position of the completed building or structure, with location of the project central to the construction project process. However, this is now changing through the use of offsite manufacture and assembly techniques.

If we consider the role of location in the procurement practices and supply chains or clusters, we can see this has many direct consequences. For example, historically shipbuilding dominated specific regions of the UK, with the local economies being heavily dependent upon the fortunes of the local shipyards not only for directly employing many people, but for placing numerous orders with other suppliers that would cluster near. As the fortunes of the UK shipbuilding industry began to wane as preferable suppliers appeared overseas, so the levels of direct and indirect employment began to fall.

In a rather different sector, the UK film and TV production industries are potentially less dependent upon location; whilst major films require huge sound stages, smaller studios are capable of being created more quickly and easily. Therefore film and TV production, when considered as projects, can either be dominated by the need for specialist production facilities or generated entirely 'on-location' if there is sufficient justification. The south-east of England dominates such project-based production, with large studios to be found in Hertfordshire and the north of London. In this locale are also a large number of specialist suppliers, providing a range of expertise either for the production itself, or in the increasingly important post-production phase. When there is the need for the production activity to take place on-location, there is a very significant logistical exercise to move all the various film-producing elements to the location specified. The associated costs involved are potentially large, therefore prompting great concern about managing the schedule and dealing with uncertainties and risks such as weather and health and safety (DeFillippi and Arthur, 1998).

Part B

Construction, in contrast, has always been accustomed to being based at the site where the construction is taking place, although design and some other consultancy services is often completed in other locations during the course of the initial problem definition and information development stages. This locational freedom for the professional services fraternity has allowed some to become international and even global players. This is particularly the case for architects, where the best are either sought by commission or enter through competition on a global stage. Linked to architecture, the engineering challenges that arise from the design intent, the relevant geographical/geological/topographical issues, or the applicable regulations have also led to the development of skills that are demanded across national boundaries. This design and consultancy skill base is completed by those specialists that focus on aspects of cost, risk, or project management, on occasion exporting a complete national approach to a foreign country as part of historical connections or as specified as part of the client's or project financier's requirement. The majority of construction in the UK is, however, anchored in the practices conducted by national players. There is also significant regionalisation, with very many construction companies having clearly defined epicentres of operation and limits of the geographical coverage.

The modern UK construction company is likely (although there will always be exceptions) to be a part-player in the larger construction process. The construction market is no longer dominated by cradle-to-grave suppliers capable of providing all the necessary services, and it was mainly house builders and road builders that had the means, motive, and opportunity to devise, design, and deliver completed projects using totally in-house resources (Hillebrandt and Cannon, 1990). Construction in the UK is now highly fragmented, with the largest classification of firms (employing over 1,200) being only 0.03% of the total number of firms and these 60 largest companies being responsible for only 19.1% of total output in 2007 (ONS, 2008–Table 3.1). This fragmentation requires sophisticated structural mechanisms to join the large number of players together to generate the value-added output that clients require. This has led to the development of numerous contract types, sophisticated project management systems and techniques, and the focus upon supply chains and their management. This last point is the focus for this chapter and is set in the context of the risk that is present when a collection of project actors are linked to achieve the delivery of a project for a client.

A modern construction project in the UK is therefore not capable of being reduced to a standard definition. What is likely is that such a project will involve a number of organisations that interact with each other to deliver the client's expectations. Figure 6.2 illustrates this point.

Simple construction projects may involve a few tiers of relationships with few players in each tier. Complex projects may have multiple layers of suppliers involving very many players in many organisational types – from the international and publicly listed to sole traders. In both simple and complex projects, it is feasible that considered chains or clusters of suppliers may be required to work with each other, either pre-aligned through strategic alliances / long-term partnering agreement, or forced through project-specific competitive tender.

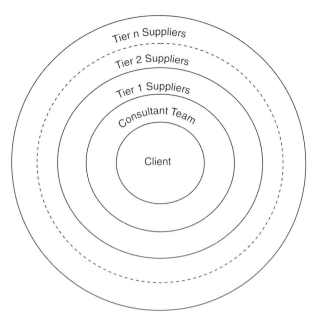

Figure 6.2 The tiered supply chain.

6.3   Supply Chains and Risk

The term 'supply chain' may suggest a coherent and interlinked set of supply relationships that at the point of ultimate delivery provide a complete solution. This is the case described in *The Machine That Changed The World* (Womack, *et al.*, 1990). On complex projects such supply chains (involving many external suppliers) are essential to respond to the size and scope of the packages of work that are proposed by those in the position to act on behalf of the client. The management of the integrated supply chain offers the opportunity to capture increased value and to minimise the risk to the client. There are many uncertainties and risks involved in construction and whilst some of these risks can be entirely negative, for example, a task being potentially physically dangerous, many are of a commercial nature, with the impact of a risk event being detrimental to cost or profitability, dependent upon the individual's position in the supply chain. Construction in the UK has become fragmented as a result of the need for flexibility and the uncertainty of future demand (Ive and Gruneberg, 2000). Large companies need large volumes of work to keep their resource base content, whereas smaller companies need less and are able to change velocity, (i.e. speed *and* direction), more easily. This has proved to be a dominant view and the number of contracting firms in the UK industry has remained large and reasonably stable, as indicated by Figure 6.3 which shows the number of privately owned construction contractors registered in the UK since 1995.

If we consider that all firms fill particular market opportunities – some being multi-faceted, others highly specialist – with operational remits from

Part B

Part B

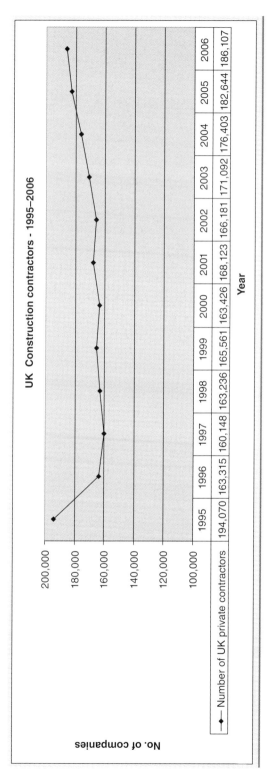

	1995	1996	1997	1998	1999	2000	2001	2002	2003	2004	2005	2006
Number of UK private contractors	194,070	163,315	160,148	163,236	165,561	163,426	168,123	166,181	171,092	176,403	182,644	186,107

Figure 6.3   Number of UK construction contractors. Source: adapted from BERR (2007), Table 3.1.

Figure 6.4   The building commissioned by Swiss Re at 30 St Mary Axe, London. Image courtesy of Stephen Pryke.

Part B

the local to the global, it should not be to anyone's surprise that managing projects will involve managing a diverse set of relationships, with the opportunity to work repeatedly with the same individuals from the same organizations being the exception rather than the rule. Major projects, such as the construction of Heathrow Terminal 5, do have the duration and size that allow for the establishment of frameworks of long term business relationships, with opportunities for learning and development at both the individual and organisational level. However, the very nature of projects means there appears little point to post-project-based learning sessions when the chances are that the project in question will not be repeated. This complexity of players associating with projects on a piecemeal basis is an accepted consequence of the macro-flexibility of the UK construction industry. It does, however, mean that significant risks, measured both in terms of potential impact and frequency of occurrence, exist on any project as organizations and individuals have to understand the risk associated with a particular project and the implications of risks across the projects that comprise the workload of the firm.

The presence of these risks requires suitable management. The way these risks are managed depends on many factors, but broadly they can be divided between those that are related to the project and those that are related to the players' perceptions, which are driven by both level and type of education and previous experiences (Maytorena *et al.*, 2007; Chapman, 1998).

To separate out the difference consider the building commissioned by the Swiss Re insurance company in the City of London (see Figure 6.4). Amongst many complexities, the design, manufacture, and installation of the compound curving façade required the solving of many problems never previously encountered. These ranged from the development of a sophisticated 3D design model to the site-based requirement to establish near perfect datum points in a building without an obvious set of standard setting-out grid references. These can be considered as objective project risks that anyone, regardless of their own view, would need to overcome. On top of these, there were many perceived risks that would add to or build upon these objective risks. Excellence in managing projects requires both types of risk to be considered and whilst many of these risks will be discrete they will normally occur across a relationship boundary.

6.4   Supply Relationships

So far, this chapter has considered a number of key elements that bear on the topic of supply chains. We have looked at the risks and uncertainties, noted the role of an individual company in a complex project, and suggested that managing these challenges is neither easy nor straightforward. To build on this base, we now turn to the nature of the relationships that are formed as a consequence of the project challenge. Sanderson and Cox (2008) suggest four different forms of interaction with suppliers, as illustrated in Figure 6.5, with SCM being appropriate when the intended way of working is close and affirmative.

The four cell descriptors in Figure 6.5 convey the range of supply relationships that are encountered, depending on both the types of suppliers and the relationship with them. So, for 'off-the-shelf' items it is the sourcing of the item that is important (top left). This is contrasted with core suppliers

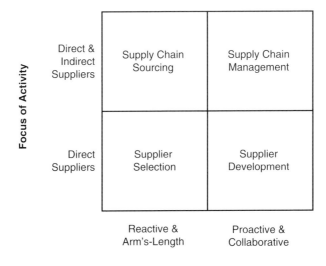

Figure 6.5 The four different forms of supplier interaction. Source: Sanderson, unpublished.

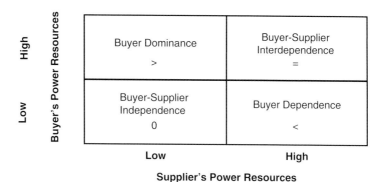

Figure 6.6 Buyer-seller power relationships. Source: adapted from Cox *et al.* (2000, p.18), with kind permission from Earlsgate Press.

who significantly interact with the procurer (e.g. a car manufacturer with many of its suppliers), where the supplier relationship is continually developed. SCM in this context is where the procurer is involved with the supply chain's actions and where that supply chain is constructed from various entities, some direct, others more disconnected.

It would appear reasonable to think that when clients or their agents select the group of primary and associated suppliers that will be charged with delivery of a project, there will be proactive and close working. This prompts the expectation that an SCM approach will be used (top right quadrant in Figure 6.5). However, as Figure 6.5 makes clear, not all relationships will be proactive and collaborative, some supply relationships may simply involve arm's length and essentially reactive transacting.

Following this logic of the appropriateness of SCM for projects, Cox (2006) considers the relative power resources (or *leverage*) of the parties. This is shown in Figure 6.6 below.

In Figure 6.6 the following definitions apply:

- '> Buyer Dominance' – Significant value appropriation for the buyer and potentially significant opportunities for SCM. The buyer has power and potential influence over the supplier.
- '< Buyer Dependence' – Limited value appropriation for the buyer and with limited opportunities for SCM. The supplier is powerful and has no reason to listen to the buyer.
- '0 Independence' – Significant value appropriation for the buyer, but limited opportunities for SCM as there is little reason for the supplier to interact.
- '= Interdependence' – Here there is shared value appropriation and significant opportunities for SCM.

The implication from the Cox hypothesis is that SCM becomes dependent on either buyer persuasion (centrally), or where benefits accrue for two firms, located in lower tiers, they might experience mutual benefits (decentralised). This can be considered in terms of the economic cycle. When the market is buoyant and order books are full or nearly so, it seems common sense that suppliers will have very few incentives to waiver from their standard offering (both in terms of supply and contract), unless the buyer has some form of extraordinary influence. To avoid this variability, some client organisations have elected to form framework agreements or strategic alliances that provide relationships that are maintained for longer than a single project, with the aim of developing a true SCM approach.

Cox's work helps to define SCM and it offers alternatives that cover a range of different supplier relationships. As such it can be considered as a final piece in the jigsaw of pieces that comprise risk and supply chain management.

It is now appropriate to bring these pieces together by considering the role that risk management plays in projects and how this is interwoven with the way that suppliers are handled, through SCM.

6.5   Risk and Supply

Construction projects are amongst the most diverse and challenging range of those in which mankind is involved. Individually, projects in other sectors such as petrochemicals, aerospace or pharmaceuticals may be more technically complex or bigger, but it is the range of construction projects in terms of number, size and complexity that makes them highly significant. As noted previously, construction in the UK has adopted a plethora of contractual forms and devices to cope with this vast range and great variability. The tide of contract types has ebbed and flowed, with times when there was consolidation around a few key types, to other periods when there was very significant choice as to which form to use (Masterman, 2002). The introduction of the New Engineering Contract (NEC) is telling. The first version of this new type of contract in the 1990s elected to use the word 'New' in its title. Time has eroded the impact of such a classification and the danger of using a word such as new is that it begs the question: what will follow? The answer (as of 2007/8) is that the original NEC has now evolved into NEC3 under the guise of the Engineering and Construction Contract (ECC). So, the *New* Engineering Contract is no longer new and it has been superseded more than once, thus demonstrating how the needs of the contract will continually change over time.

A fuller investigation into the purpose of contracts is beyond the limited scope of this chapter, but it should be remembered that contracts are used to establish expectations, roles, responsibilities, provide incentives and penalties and are enforceable through independent third parties. From a management perspective they are a key risk management device. If a SCM approach is taken as the basis for the management of the construction project supply relationship, then it becomes clear that there are two potential risk management routes: use the contract or manage the relationship.

The choice for those responsible for managing projects may be considered as the development of strategies to pursue specific performance or financial compensation through the contract, or to adopt a relationship management approach (Pryke and Smyth, 2006) to achieve the same objectives. If the parties are clear that the contract is to be the focus of project governance and control, then a typical management plan would be to mitigate through the use of contingency. Thus, parties will use contractual terms in a back-to-back form, making sure that obligations they receive in their contract are passed on to those that they, in turn, will contract with. Deviations from this occur at points where the 'risk' in question is clearly more effectively handled in a different way than simply passing it on to the next tier of contracting party (refer to figure 6.2, page 121). This may be that the risk is eliminated within the organisation through the deployment of internal expertise and other specialist resource. Alternatively, an insurance policy, or some other form of risk transfer, may be used. The danger here is that either through naivety, ignorance or incompetence risks are accepted for which there is inadequate ability to manage. Take for example a brown field site redevelopment: If a contract is awarded for clearance of the site (to include remediation and decontamination) then a relevant question arises: 'What is in the ground and what was the land previously used for?' There are many ways to answer this, ranging from the 'don't really know, but after a casual inspection it looks OK', to a full-scale and detailed sampling and survey exercise. The former approach is least expensive but may not provide the right answer, whilst the latter is more expensive in time and money and will provide substantially more data and information, but it will still require a decision as to what subsequent action to take. Deciding on the right course of action is a management responsibility and it will depend on many factors linked to the project and parties taking the decision.

British Airports Authority took the very bold move of recognising this dilemma when awarding the contracts for Terminal 5 at the UK's main airport – Heathrow (see also Chapter 8). It recognised that many possible downside problems were likely to be factored into the submissions made by contractors tendering for work. This 'pessimism-bias' was going to be at BAA's expense and would create a mindset of focusing on the problem, rather than seeking to avoid problems in the first place. BAA therefore elected to hold the risk that there would be problems, and ensured that such problems were paid for out of project held risk funds. Failure to spend this project risk fund would lead to unspent residual monies flowing back to the contractors. This enabled a series of incentives to be offered to the contractors to seek to eliminate problems that may otherwise have incurred additional time and costs.

Whilst there are those that will always prefer contractual redress such initiatives as used by BAA on Terminal 5 contribute to our understanding of complex inter-firm relationships that might be effectively managed through SCM in the project environment.

6.6   Managing Risks in the Supply Chain

When considering the project challenge and the management of that supply chain as a route to project success, there are two important viewpoints: the project perspective and that of the participating supply chain organisation (SCO).

From the project perspective, the objective is to find members of a supply chain that can work with the greatest harmony to achieve optimal results by delivering effective and efficient solutions. This will be to aim to deliver maximum benefit at minimum cost, and generating minimal waste. From the SCO's perspective, the objective is to derive maximum business benefit, measured at a base level by direct contribution to bottom-line profit. However, this short-term focus on purely monetary objectives is augmented by other levels of consideration, such as reputation enhancement, access to new markets or clients, access to new technology, or opportunities to innovate. As such the incentives may not be directly and immediately financial. However, considering the purely profit-based drivers, the implications for the SCO are that it must carry out the necessary work in the most effective and efficient way to allow the maximum sales revenue to be achieved. We therefore see some degree of symmetry between the needs of the project and those of the SCO. An optimal solution would be where the project's needs are matched exactly by the SCO's offering, whether that is for bespoke products or services, or more off-the-shelf solutions.

The concern about risk management in such a context relates to two issues: time and information. Time is important as SCOs are selected in anticipation of what they are capable of, based on either direct bidding, or some form of recommendation based on past performance or referral. The issue of information arises as the link between project need and SCO is made on the basis of the available information. Both these issues are of interest to those who study Transaction Cost Economics (TCE), where issues of timing and information asymmetry are part of the explanation for what are observed as being explanatory factors behind economic decisions (Williamson, 1975, 1985).

For those not particularly interested in TCE there is a link derived, arguably between time and information, with the expectation that as time passes, so more information is potentially available (Winch, 2002). The SCO should be trying to get appointed to those projects that allow it to maximise the benefits available, by doing the work required faultlessly and with minimum effort and resource. Whilst this might be the aspiration of both sides, the reality is that wrong decisions are made, with either one or both sides losing out significantly, sometimes with dramatic consequences.

If this problem is cast in the light of risk management, it becomes one of management failure. This can be deliberate: for example, project clients wanting to have certain companies working on a project because of the kudos they bring, irrespective of their appropriateness; or SCOs deliberately setting out to win contracts at below commercially viable terms or beyond their technical competences because of other commercial (e.g. cash flow) or non-commercial (e.g. market entry, reputation) reasons.

The principle that 'risk is placed with those best placed to handle it' is illustrated by the following quotation taken from the Her Majesty's Treasury in the UK:

'The governing principle is that risk should be allocated to whichever party from the public or private sector is best placed to manage it. The optimal allocation of risk, rather than maximising risk transfer, is the objective, and is vital to ensuring that the best solution is found.' (Source: HM Treasury Green Book, Annex 4, Section 1.3 http://greenbook.treasury.gov.uk/annex04.htm, accessed March 2008)

When considering the above quotation, the question of how to define 'optimal' becomes of interest. This is because managing risk is always accompanied by a cost. In some cases, this cost will be real, with parties only agreeing to do something upon receipt of payment, through to the purchasing of insurance policies to compensate, should the risk occur.

An example of the difficulty of establishing accurate compensation for risk, which is difficult to accurately quantify, involves the opening sequence of the James Bond film 'The Spy Who Loved Me', where in 1976 the stuntman (Rick Sylvester) was paid $30,000 to perform a stunt involving skiing off a cliff and then detaching his skis and opening a parachute. This stunt was considered uninsurable as there were no precedents and many potential problems and dangers, so the fee was paid to the stuntman to reflect his personal assessment of the potential risk. The outcome was a successful stunt at a personal and professional level for the stuntman and team, and a notable piece of film history that cemented the James Bond brand. This example illustrates how risk must be valued by those unable to transfer risk to others, even where the extent of potential risk and the implications are different or impossible to quantify.

There has been academic investigation into what John Adams refers to as our individual risk thermostats (Adams, 1995), which suggests that each of us (individually and organisationally) has an inherent tolerance for taking on risks. This explains why some people gamble (and the limits they set on the stakes they are prepared to commit as bets), as well as explaining why organisations vary in their risk appetite. Whether individually or organisationally, there is a spectrum on which we all sit that ranges from great risk adversity to enthusiastic acceptance of risk. Indeed, it can be argued that individuals and organisations can appear almost schizophrenic, being in some situations highly risk adverse (e.g. financial investments), whilst in relation to other unrelated issues being risk seeking (e.g. taking on untested new recruits). This reinforces the thrill/threat dichotomy that both excites and frightens in equal measure.

The question of drivers that influence the response to risk-laden situations or opportunities leads to the consideration of rational decision making, where all the relevant factors are identified, weighted according to the decision maker's personal risk perception, and then used to make the decision. Whilst it may be the case that a third-party observer cannot understand why a particular decision was taken, it remains that the party making the decision does so recognising the possible consequences. Society, and indeed significant parts of the judicial system, are there to reinforce the serious nature of the wider consequences of decisions are not made carefully.

An example are the laws against drinking alcohol and driving, where, in the UK, a relatively small degree of personal discretion is available, but serious penalties are in place for those that transgress. These penalties exist because of the wider ramifications that exist to others who had no part on the drink-drive decision, but are affected by the potentially tragic consequences when motor vehicle drivers under the excessive influence of alcohol cause accidents or incidents that hurt others.

These factors are at play in the project supply chain context, with parties accepting risk being explicitly aware of what the risk is, seeking to take that risk at a cost to the party seeking to transfer it, and then working to eliminate or solve it. This was one of the cornerstone arguments for the use of the Private Finance Initiative (PFI), where the UK public sector had an indisputable historic problem of sanctioning projects that involved the construction of built environment assets, only to find that outturn costs and dates were far greater and longer than anticipated. The effect of using PFI as a procurement method is to place relatively large penalties on the service provider for failure to comply. The result, arguably, has been an improvement in delivery performance of construction projects, with those responsible for delivering the project utilising best practice in many areas to ensure that expected results are achieved.

6.7   Supply Chain Management and PFI

The development of PFI and its increasingly global influence has, to a limited extent, shifted the debate about risks and their management from certainty of delivery in terms of cost and time, to the risk of damage to the visual urban landscape of built assets that have (in a number of cases) very limited aesthetic appeal. This is a good example of how a lack of accurate risk identification can produce undesirable consequences. In the case of early PFI projects, the challenge was clearly set by the public sector project client for effective and efficient delivery of built assets that would underpin the delivery of a service. The private sector actors responded with standing supply chains – established and maintained over medium term periods, typically via 'with design' forms of contract and transfer of contractual obligations down the supply chain. Lean and agile approaches were adopted and modern methods of construction technologies involving modular assembly and off-site manufacture heralded a new era of construction delivery certainty. The aesthetic appeal of the design was often reduced to a lower weighted priority that was capable of being sacrificed for improved certainty of delivery and functional performance.

The criticism raised by commentators of architectural impact gathered pace as more products of PFI appeared and it was appreciated that there is a value at the urban scale accruing to good design. Latterly, PFI projects have, therefore, had to achieve the additional requirement of appeasement of architectural critics, along with certainty of delivery – clearly an overall increase of risks. The solution has been to focus on the designers within the supply chain and challenge them to produce design solutions that are not only straightforward to deliver and operate, but also capable of impressing a critical set of design evaluators. Whether the PFI project was in the era before aesthetic concerns were elevated or before, those companies responsible for delivering the solution needed to have some sort of strategy towards design and management. In the former case, it was not for the suppliers to worry about the appeal of the design, whereas in the latter case it was. But what of those who happened to be working on projects in the transition phase where this concern about visual appeal was raised as the design was still being developed? The organisations in this situation would have felt they were facing an unexpected challenge (risk) that they may not have previously considered. This is an example of the opposite situation to that outlined in the quotation provided earlier from H.M. Treasury. Far from the risk being handled by those best placed to do so, the risk eventually falls upon the party that is least able to protect themselves from risk transfer.

There may be those who, at this point, would introduce the concept of opportunism (Williamson, 1975). Instead of unambiguous, neutral and objective information being given by one party to another, incorrect, untrue, or partial information is provided, resulting in an agreement being reached that later proves to be wrong for at least one party (Winch, 2002). This information asymmetry may be with or without malice.

Academics have for some years enjoyed working with game theory and there is now a body of sophisticated mathematical models that illustrate particular scenarios or games, which have been tested empirically and found to be valid. A notable example is the UK government bidding competition for the broadcast licences for third generation mobile phone networks. The rules of the auction for the licences were created using game theory modelling, with a result that bidders worked with information available and their assumptions about others' actions and future developments to generate a bid sum. The result was the submission of extraordinarily large sums of money that benefited the UK Exchequer, but that will take years to recoup by those that 'successfully' won the auction.

The emerging picture is one where project supply chains are managing a number of risks and opportunities; some of these risks and opportunities are easily identified and managed – others less so. If we set aside criminality, which can totally dominate all other factors, we have the fundamental challenge of finding project suppliers that are best placed to solve the particular project challenge. Whilst there may be many possible providers of a solution such as to provide the architectural design or act as the principal contractor, it is highly unlikely that the choice set will be totally unconstrained, with the actual choice being limited by factors such as geography, timing, resource availability, previous experience or other specific requirements for a competency, skill or technology. These drivers bring into play many other factors that begin to alter the reality from simple 'risk management'. These other factors include: the relative power bases in the supply chain (see Figure 6.6, page 125); and the propensity to ensure maximal optimality, requiring the greatest effort to consider *all* options as against satisficing, requiring the least effort that is acceptable.

When the word 'power' is used in this context it should be defined carefully (Cox *et al.*, 2006). Power to make decisions or influence decisions can come from many sources. Financial power is clear, as those with significant financial backing are able to resist altering their position for longer than those with more limited financial resources. This may be as true of those in debt as of those with substantial resources and able to pursue legal redress. Political power is also important, whether this is at the organisational or individual level. As we move away from tangible to more intangible sources of power, so we need to consider factors like knowledge and reputation. As sources of power become more complex to understand and measure, so the management challenge becomes more daunting, as decision making occurs at points in time and based on certain factors known at that time. Not all the factors may be clear, nor the full effects of the decision understood.

When combined, as the issues listed above will inevitably be, there can be a complex cocktail of factors at play when allocating risks in project supply chains. Managers of projects therefore value experiential learning very highly, as the degree of nuance and idiosyncrasy at play in a real project can mean a vast difference between the obvious factors and those operating at a subtler level. For this reason, those who manage projects, risks or supply chains learn to constantly adapt their approach. How much alteration and adaptation that is needed is a function of the degree of variability of the projects they work on and who they work with. Hence, there has been the recognition that partnering, alliancing, or working with prescribed frameworks has significant advantage because this takes away the need for reactive alteration, instead allowing for proactive and planned alteration to managing the project. The supporting arguments are that the projects benefit, being operated from a starting point of familiarity in process and design, the organisations can afford to invest in becoming compatible through lower transaction costs, increased 'numbers', and individuals can feel more confident in the approach they will take and the response they will evoke. The positive outcomes are potentially more successful projects, increased benefits to the organisations involved and greater satisfaction for the individuals. The negatives are that either the positives are not achieved, or that complacency sets in.

Regardless of whether a project is isolated or part of a suite of projects forming either a programme or a disparate portfolio of unrelated projects, the challenge for managing risk within the project supply chain will be continuous. The same risk may be viewed by some as a problem or threat, whilst others simply adopt an entrepreneurial approach and view the transfer of risk to their organisation as an opportunity to increase workload and profits. Take, for example, the challenge of cleaning glass atria that have become a common feature for many office-type buildings. Whilst it is perfectly legal for the designers to assume that such structures will only be capable of being kept clean by use of individuals using rope access techniques derived from mountain climbing, there is the practical challenge for building owners and operators to find companies willing to deliver such service and which allow for the relevant insurance cover to be provided.

Fortunately, there are brave individuals who, presumably, relish the prospect of cleaning and maintaining elements of buildings whilst hanging high above the ground. The prevalence of such companies and their track record of safety and efficacy clearly indicate the willingness of some individuals to specialize in certain areas of risk, to manage it effectively and to profit from such risk.

Wherever a project client seeks external organisations to deliver the project solution there is an example of risk transfer. When this works well, the parties accepting the project challenge know clearly their span of skill, resources and competences. They will absorb those that they are best placed to resolve, transfer to others those that are in the same relative position as themselves, and reject others that they cannot cope with. The risks in this last group can be the ones that end up providing the greatest source of discussion, debate, and disagreement as they are typically perceived as ominous threats. Whilst there is a very wide range of examples, some of the most challenging can be considered as *force majeure* (greater force), where events or other manifestations beyond the control of either contracting party thwart the delivery of the project. The global threat of terrorism is something beyond many projects' ability to influence and such an act carried out on a project would not be seen as being the fault of one party. This may appear reasonably obvious and common sense, but it will be in a contracting party's interest for commercial self-preservation to try and get exoneration for any risks that it cannot accept or transfer. Clearly, the other party to the contract will have incentives to argue against that view stating that there may be mitigations available.

A traditional British preoccupation is with the weather, which can, and has, led to many risk management based discussions on the nature of risk associated with weather and more recently climate change; which party is bearing that risk, and what to do if some particular weather eventuality occurs. Where water damage to exposed building interiors is a possibility, the pattern of rainfall in the UK would make it almost impossible to be certain that rain will not fall on any exact location at any particular non-imminent future time. How to handle this risk is, therefore, a matter of debate and judgement. Protecting sensitive work from potential rainfall can be expensive and time consuming, but if you bear the risk for the consequences then you may feel you have no choice. If, however, there is some way of predicting weather better than others then you may be able to win more work by offering to carry out the works without any such protection. Such risk taking will be successful as long as your knowledge and decisions are correct and accurate, but until we understand more about weather patterns, it is unlikely that many in the UK will see it as being anything other than random. The concerns about greenhouse gas emissions may have direct bearing on this as there is now considerable resource dedicated to understanding their impact on weather patterns. It is too early to say whether climatic models will prove to be robust and accurate at the scale needed to influence projects, but such knowledge generation will be available for those assessing this risk on projects and may influence their decision whether to retain or transfer that risk.

6.8   Concluding Remarks

The delivery of a project through the use of SCM is increasingly becoming the norm for many industries and sectors. Utilising the range of skills, competences, experiences and knowledge bases of the organisations and individuals involved may offer the project a tremendous opportunity to produce excellent project outcomes. Management of any supply chain is anchored in the management of risk that both the project and the supply chain represent. This is very much about upside risk as well as downside, with potential good ideas and unexpected opportunities being present, as well as the problems and threats to effective project delivery.

In this chapter we have considered that risk will flow through the supply chain and have raised a challenge to those who simply state that risk will reside with those best placed to handle it. It is recognised that, as individuals and through the organisations that we work in, we are more sensitive and proactive about risks presenting a threat, than uncertainties representing potential benefit. Major projects tend to fail to meet objectives more than they surpass them. This prompts the investigation into whether the problem is caused by a failure in the supply chain to either appreciate or defend against the allocation of inappropriate risk. This leads to issues associated with information flow and relative power of supply chain actors.

Where the critical information does not exist or is not made available there will be opportunities for assumption or ignorance to exist, both of which can lead to downstream consequences from upstream decisions. All projects are conducted within the context of contractual conditions and it is possible that whilst there may be sympathy for problems caused from unanticipated problems, contracts are in place to apportion responsibility and offer routes for redress. When resort to the contract is needed it can be a protracted and costly exercise, with such significant ramifications as to take the focus off the project and on to the dispute, detracting from the strength of the arguments made above.

The consideration of power in the supply chain is one that has been investigated and which can be viewed in a number of ways. Market power is dominated by the economic interaction of buyers and sellers; therefore it moves in time and is always in a state of flux. For a project, its specific needs are likely to be fixed both in terms of the nature of the need and the timing of that need. This means that macro-market forces may become significantly irrelevant as will be the specific conditions at that time, in that locale, and in that specialist set of markets relevant to the particular project environment. Trying to militate against this dynamic variable leads to the possibility of creating longer term relationships that try to smooth out the turbulence of project-by-project procurement. Political power is more esoteric, depending upon many background factors that need to be understood in order for the relationship to be managed effectively. Political influence may be entwined with the power which comes from knowledge that is not freely shared. This knowledge power is very much an emergent field as projects present both great opportunities and challenges to extract and use knowledge. Whilst this may be understood, how this risk is managed is not yet clear.

Organisations forming part of a supply chain will be assessing, preparing, managing, and anticipating their involvement in a project either in series or in parallel. Some projects will present opportunities or threats to organisations that could make or break them. For other organisations, their involvement in multiple projects at any one time means there is no one clear signal being received about the risk management issues present, but a cacophony of risk management 'noise' that the organisation has to filter carefully and choose its response to. Whilst each organisation may have the aspiration to only accept those risks it is best suited to cope with, there may be complex reasons why this is not the case and the expected outcome for both the project and the organisations involved is either not achieved at all or in the way they anticipated.

This chapter has identified risk in the conceptual framework of the project supply chain. The UK construction industry in the past has adopted a highly adversarial approach to the management of, and compensation for, fair and unfair risk transfer occurring through procurement and project management processes. More recently it has begun in places to adopt a collaborative approach to the elimination and management of project risks.

Anecdotal evidence continues to suggest that risk is being managed by those with the least power or leverage and that more emphasis is placed on uncertainty associated with downside threats, rather than upside opportunities. This suggests that the management of opportunities within the supply chain need more effective consideration and management.

Part B

References

Adams, J. (1995) *Risk: the Policy Implications of Risk Compensation and Plural Rationalities*. London, UCL Press.

Bernstein, P. (1996) *Against the Gods: the Remarkable Story of Risk*, New York, John Wiley & Sons.

BERR – Department for Business, Environment and Regulatory Reform (2007) *Construction Statistics Annual 2007*, Norwich, The Stationary Office.

Chapman, C. and Ward, S. (2003) *Project Risk Management: Processes, Techniques, and Insights*, Chichester, Wiley.

Chapman, R. (1998) The effectiveness of working group risk identification and assessment techniques. *International Journal of Project Management*, 16, 333.

Constructing Excellence (undated). Supply chain management fact sheet, available at http://www.constructingexcellence.org.uk/pdf/fact_sheet/supplychain.pdf accessed January 2008.

Coase, R. (1937) The Nature of the Firm. *Economica*, 16(4), 386–405.

Cox, A. (2006) 'Value for whom? Win-win and the problem of interests in buyer and supplier exchange', *Proceedings of the Fourth Worldwide Symposium in Purchasing and Supply Chain Management* (Supply Chain Management Institute, University of San Diego, 6th–8th April, 2006), pp. 25–38.

Cox, A., Ireland, P., Lonsdale, C., Sanderson, J. and Watson, G. (2003) *Supply Chain Management: A Guide to Best Practice*. London, FT Prentice Hall.

Cox, A., Sanderson, J. and Watson, G. (2000) *Power Regimes: Mapping the DNA of Business and Supply Chain Relationships*. Stratford-upon-Avon, Earlsgate Press.

Cox, A., Watson, G., Lonsdale, C. and Sanderson, J. (2006) Strategic Supply Chain Management: the Power of Incentives. in Waters, D. (ed.), *Global Logistics: New Directions in Supply Chain Management*. London, Kogan Page.

DeFillippi, R. and Arthur, M. (1998) Paradox in Project-Based Enterprise: The Case of Film Making,, *California Management Review*, 40(2), 125–139.

Flyvbjerg, B., Bruzelius, N. and Rothengatter, W. (2003) *Megaprojects and risk: an anatomy of ambition*. Cambridge, Cambridge University Press.

Harvard Business Review (1999) *Managing Uncertainty*, Boston, Mass, HBS Press.

Hillebrandt, P. (2000) Economic theory and the construction, Basingstoke, Macmillan.

Hillebrandt, P. and Cannon, J. (1990) *The Modern Construction Firm*, London, Macmillan.

Ive, G. and Gruneberg, S. (2000) The *economics of the modern construction sector*, Basingstoke, Macmillan.

Keynes, J. (1921) *A treatise on probability*, London, Macmillan.

Knight, F. (1921) *Risk, Uncertainty, and Profit*. Boston, MA: Hart, Houghton Mifflin Company.

Masterman, J. (2002) *An introduction to Building Procurement Systems*. London, Spon.

Maytorena, E., Winch, G. M., Freeman, J. and Kiely, T., The Influence of experience and information search styles on project risk identification erformance. *Engineering Management, IEEE Transactions on*, 54(2), 315–326.

Miller, R. and Lessard, D. (Eds) (2000) *The Strategic Management of Large Engineering Projects*. Cambridge, Mass, MIT Press.

(ONS) Office of Natural Statistics (2008) Construction Statistics Annual, Basingstoke, Palgrave Macmillan.

[Presidential Commission] Report of the Presidential Commission on the Space Shuttle Challenger Accident, (1987), Chapter 5, [82]). Available at http://history.nasa.gov/rogersrep/v1ch5.htm (access December, 2007).

Pryke, S. and Smyth, H. (eds) (2006) *The Management of Complex Projects: A Relationship Approach*. Oxford, Blackwell.

Sanderson, J. and Cox, A. (2008) The challenges of supply strategy selection in a project environment: evidence from UK naval shipbuilding. *Supply Chain Management: An International Journal*, Vol. 13 (1), 16–25.

Smith, N. (1995) *Engineering Project Management*. Cambridge, Mass, Blackwell Science.

Williamson, O. (1985) *The Economic Institutions of Capitalism: Firms, Markets, Relational Contracting*, New York, Free Press.

Williamson, O. (1975) *Markets and Hierarchies: analysis and antitrust implications: a study in the economics of internal organization*. New York, Free Press.

Winch, G. (2002) *Managing Construction Projects: an information processing approach*. Oxford, Blackwell.

Womack, J., Jones, D. and Roos, D. (1990) *The Machine that Changed the World: based on the Massachusetts Institute of Technology 5-million dollar 5-year study on the future of the automobile*. New York, Rawson Associates.

Part B

7

Slough Estates in the 1990s – Client Driven SCM

Bernard Rimmer

7.1   Introduction

This chapter describes the experience of a hands-on property company in a period of its history when senior management took an unusually strong interest in design and construction, and positively encouraged the application of innovative approaches to both design and procurement of its commercial developments. The Chairman of Slough Estates at the time, Sir Nigel Mobbs, was always pointing out the lower costs of industrial units in the other parts of the world in which the company operated and was keen to back any ideas that had potential for changing UK construction performance. The author, as head of the design and construction team, responded to this challenge and, with the freedom he was given to experiment, went on to pioneer improvements to the ways Slough Estates buildings were procured. Costs indeed were lowered, but never to the '£ for $' level of the USA.

At the time this was all happening, supply chain management (SCM) was only just being talked about in the construction industry and Sir John Egan brought concepts such as lean supply to its attention (Egan, 1998). Nevertheless, a great deal of what was being done at Slough Estates was SCM-type activities, although it was not classified or described as such within the organisation at the time. Much of the text in this chapter is expressed in terms that hopefully will contribute to the body of knowledge of SCM practices in construction, while acknowledging that there was never an intention at the time to pursue such a result.

The chapter begins with the definition and scope of SCM as the author sees it applying to construction, and goes on to analyse the main UK procurement systems in terms of the potential for each in the application of SCM principles. This is followed by a summary of the main influences on company practices of the two major reports commissioned by the government on the functioning of the construction industry: Latham (Latham, 1994) and Egan (Egan, 1998).

Set against this background of UK procurement turmoil and the exposure to new ideas from other industries, the main body of the chapter is a description of what happened in practice at Slough Estates in the 1990s. The attitude of senior management and the organisation's structure are discussed in detail to emphasise the importance of leadership in attempting to apply SCM and lean principles to construction projects. Key initiatives that were taken are described in detail, and comparative results from benchmarking other projects are presented.

The final conclusions summarise the lessons learned from Slough Estates' innovative period in the 1990s, and possible limitations in the future development of SCM in construction are discussed in light of the general reluctance of client organisations to maintain close involvement with the construction process – essentially their reluctance to participate in the management of their supply chains.

7.1.1   Definition and scope of construction SCM

For the purposes of this chapter the following definition and scope of SCM will be adopted, recognising that there are other ways of defining this 'autonomous managerial concept' (Vrijhoef and Koskela, 1999).

7.1.2   SCM definition

The management of upstream and downstream relationships with clients and suppliers to achieve greater project value at less cost.

This is a modified version of a widely accepted definition (Christopher, 1992) which takes account of:

- the bespoke, one-off nature of construction projects;
- the fact that designers, being largely independent firms, can be regarded as suppliers;
- the fact that the client is a key actor in the process.

The definition would imply that SCM in construction is the prerogative of main contractors but, as will be seen later, hands-on clients themselves, such as developers, can take the lead role in SCM. In this case, the upstream management of relationships can be internal (colleagues) and external (tenants, purchasers or funds). Also, the definition does not preclude SCM activities being controlled down the line; for instance, when a specialist contractor manages the relationship with the contractor (his client) and his suppliers to create better value and lower costs.

Relationships can be direct, overlain with a binding contract, or indirect, involving management communication or design liaison. However, all include soft issues such as openness and collaboration.

Value is defined as any characteristic, feature or performance (of a design, building or project) that is important to the customer.

Cost is the final amount paid by the client for the project, including time-related costs such as interest on borrowings.

7.1.3   SCM scope

It is clear that, in the complex and interdependent world of construction, SCM will not be effective unless it is applied by strong leadership in a systematic way over the whole supply chain in a collaborative construction environment (O'Brien, 2001). Thus, while there are important tools of SCM such as Just-in-Time (JIT) and Logistics, the main focus of this chapter is concerned with the leadership necessary to create an environment in which the various actors can work positively, proactively and enthusiastically together.

The main scope for SCM at Slough Estates was:

- A focus upon the definition and delivery of value;
- The creation of contractual arrangements in which SCM tools could flourish;
- Investment in product development;
- The search for and elimination of waste;
- Performance measurement and bench-marking.

7.2   Slough Estates' Experiences of Procurement Analysed in Terms of SCM

7.2.1   Historical background

Since its formation in 1920, Slough Estates, a property company specialising in industrial units for rent, has regarded expertise in town planning, design and construction as essential ingredients of its own business knowledge. It was second only to Trafford Park in pioneering the construction of units available for rent and in its first marketing brochure offered companies units that were 'excellently designed for works of various descriptions' (Cassell, 1991) and which were designed by its own architects and built by its own workforce. For many years, up until recently, it retained its own architects and did much of its building work in-house, although post war the work was carried out by sub-contractors.

In the 1980s the company was growing rapidly, building design was becoming more sophisticated, new market sectors such as retail and city offices were entered into, and more work needed to be procured from outside architects and contractors. Project management skills were acquired to handle outside work and a new era began, characterised by new procurement systems and increasingly sophisticated building designs. The in-house design and construction team was retained to carry out the baseload of the industrial work. The calibre of the team was improved and eventually it took on business park and office design, deservedly winning a BCO Award for its work.

During the 1980s and early 1990s the company experimented with all the mainstream methods of procurement, including traditional, management contracting and construction management. However, the results in general were unsatisfactory. Overspend was common and the company was often left with unacceptable levels of defects, often paying for rectification work. Other clients were having similar problems with project delivery and one group of clients, the British Property Federation, produced its own procurement system (British Property Federation, 1983), to which the author made a contribution. The BPF procurement system was a private sector client led attempt to reform the construction industry. Unfortunately, the system, while introducing for the first time the concept of the 'Client's Representative' (spawning the project management profession) and novation of consultants, was perceived by a majority in the construction industry as over-bureaucratic and failed to engage the industry in a collaborative fashion.

7.2.2   Procurement systems as supply chains

Each of the main procurement systems used by Slough Estates will be discussed in terms of how relationships are managed and how suitable each system is for the introduction and exploitation of SCM. This analysis helps to explain why the systems delivered poor customer value. In the diagrams for each system, the various sub-contractor descriptions and linkages are clarified as follows:

- *Specialist Contractors*: Those sub-contractors that offer a design and install service either in systems, such as curtain walling, or in engineering, such as mechanical and electrical services. Design can cover the complete service of shop drawings to consultants' designs.
- *Trade Contractors*: Those sub-contractors that have specialist skills but do not offer design services. Examples are bricklaying, groundworks and joinery.

Specialist contractors often sub-contract parts of their work to other specialists and/or trade contractors but the diagrams do not show this for simplicity. However, this cascading down the chain, sometimes with up to eight levels of sub-contracting involved, is an important feature of the industry and causes difficulties for efficient SCM.

Figures 7.1 to 7.7 deal with the most common forms of procurement and show three types of relationship existing between the project actors. These are:

- *Chains*, which represent *direct* relationships between the parties where money changes hands.
- *Strings (continuous)*, which represent *management relationships*, with no contract between the parties, but within which legally binding instructions can be given as provided for in a chain contract. An example would be an architect's instructions.

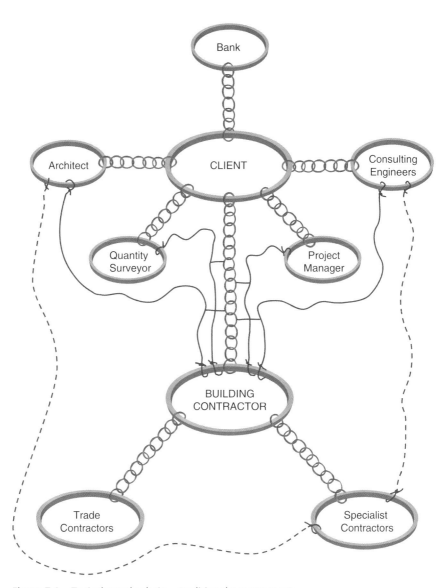

Figure 7.1   Typical supply chain – traditional procurement.

- *Strings (broken)*, which represent *design relationships*, with no contract between the parties, but within which legally binding instructions can be given as provided for in a chain contract. An example would be the engineer's approval of specialist drawings.

7.2.3   Traditional procurement

Figure 7.1 shows diagrammatically the relationships in traditional procurement.

The management of upstream relationships with clients in traditional forms of procurement is largely covered by the design team, and in particular the architect. Indeed, it is common for contractors to have little or no contact with their clients on a day-to-day basis. The architect and engineers communicate with the clients about what they want (taking the brief), then convert the brief into concepts and, ultimately, into drawings and specifications for onward transmission to the contractor. The contractor can only then pass this 'fixed' information down to the specialists and trade contractors to price, with no opportunity to establish the right environment for SCM to work. The contractor has no meaningful upstream relationship with the client to work with his downstream relationships with the main production suppliers; therefore, an in-built discontinuity is present in SCM terms. The only significant contribution that could be made by the contractor would be to offer, where the competitive framework of the contract arrangements allows, an experienced supply chain that has worked successfully with him on similar building types. Ironically, the traditional procurement model of completed designs being passed down to suppliers for price competition is similar to the process used by the British car industry prior to the Japanese arriving on the scene with Total Quality Management (TQM).

This poor situation for SCM can be improved if clients are willing to negotiate with contractors and specialists right from the start, including them in design and costing deliberations. This would effectively mean that the contractor could regard his relationship with the architect as one of the upstream relationships that needs managing in relation to value improvements and cost savings for the client. This is not as easy as in a design and build situation where the contractor has full control, but nevertheless has possibilities if the designers will collaborate. The trouble is that clients which are comfortable with the traditional form of working are usually insistent on competitive tendering, leaving few opportunities for negotiation. They are generally unaware of the huge waste in the traditional competitive process and are sacrificing potential improvements (innovation in process and product or better value, for example) for what they (mistakenly) regard as lowest cost.

It must, therefore, be concluded that the UK traditional form of procurement is a weak platform from which to add value and reduce cost through SCM, which is the reason that SCM is mostly absent from such arrangements. None of the actors, except the client, is in a position to take a leadership role and, by definition, clients who use traditional contracts are generally of a mind to trust their professionals and push all the risk 'down the line' by generally choosing to externalise risk to contractors and subcontractors through the winding of contract documents. Value engineering is one of the few tools of SCM that is used and this tends to be mainly by the design team prior to issuing tender documents.

Slough Estates turned to management contracting in the belief that benefits would accrue from having building expertise available in the design phase.

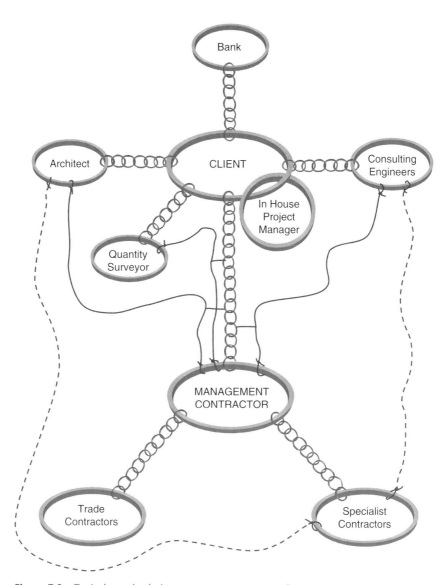

Figure 7.2   Typical supply chain – management contracting.

7.2.4   Management contracting (MC)

Figure 7.2 shows that the supply chain for this method looks similar to traditional procurement in Figure 7.1, but in principle the management contractor has a better opportunity to influence upstream relationships with the client, particularly at the crucial design phase. Unfortunately, the management contractor has no financial interest in the payments to trade and specialist contractors (the package contractors) and, based upon anecdotal evidence, has not shown much interest in managing the upstream and downstream relationships to the client's benefit. Management contractors seem to prefer to concentrate on dividing the work into clearly defined work packages and on programming issues. MC, therefore, is like a cluster of traditional projects with design discontinuity problems and split responsibilities, and is managed by a contractor with no direct interest in the financial outcome. MC also suffers from the fact that, while MC contracts are all drafted on the basis that the management contractor bears no responsibility for the performance of the package contractors, in practice contractors have been sued successfully for their part in cost and time over-runs and for their contribution to the cause of defects. MC is thus an altogether unsatisfactory platform from which to operate SCM successfully.

Not surprisingly, Slough Estates suffered from its involvement in MC. Serious problems arose on all of the projects. The company, well known in the industry for fair dealing and long-term relationships, was involved for the first time in its history in construction litigation, mainly in attempting to recover the costs of rectification of defects – the causes of which none of the parties would own.

Slough turned to construction management in the hope that direct relationships with the trade and specialist contractors, that had served the company so well on the in-house projects, would enable faster and better decisions to be made. Additionally, it was believed that the rapid payment of bills, a policy of the company, would provide encouragement to the contractors to apply their best efforts and resources to Slough projects. Unfortunately, both of these aspirations proved unrealistic in practice and Slough Estates continued its search for a SCM-friendly procurement approach.

7.2.5   Construction management (CM)

Figure 7.3 shows a typical supply chain for CM. The construction manager acts only in a consultant role and takes no contracting risk. The client takes all this risk, but direct relationships with the package contractors make the risk manageable, particularly if the client has high calibre construction executives in house. The method provides fertile ground for early involvement of the package contractors and thus opportunities are open for SCM techniques to be applied. The initiative for SCM leadership can come from the client, the construction manager or the package contractors – in their case, involving their own supply chains in the process. However, without client leadership or encouragement, SCM initiatives will be hard to promote. The best example of successful CM is that practised by Stanhope Properties, one of the most hands-on clients in the industry. Stanhope has been concentrating on customer value and driving the best design solutions using CM as its preferred system for over two decades.

The results obtained by Slough Estates using CM were much better than from using MC, but were still far from satisfactory. The company did not have a constant flow of large projects to justify the recruitment of a large in-house management team. Slough Estates had to rely on its consultant

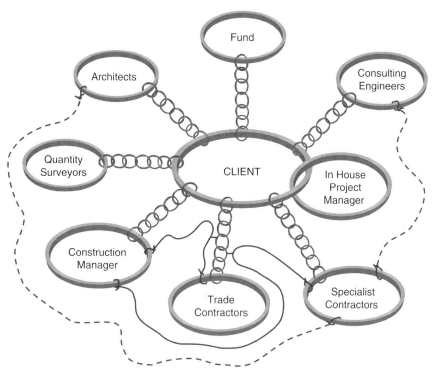

Figure 7.3 Typical supply chain – construction management.

construction managers more than its competitor Stanhope Properties, and it was felt that the best procurement solution for external projects had not yet been reached. At around the same time the University of Reading published a report (Bennett, Pothcary and Robinson, 1996) summarising the results of a research project which examined the results from over 300 projects, comparing traditional forms of procurement to design and build (D&B). Prior to this report, the common perception in the industry was that D&B in general produced shoddy results, being only suitable for small projects and industrial sheds. The study showed that while traditional projects produced slightly better quality (as perceived by customers) than D&B projects, when projects are complicated and involve high tech components the quality from D&B projects was significantly higher than from traditional methods. We now have a greater understanding of why this should be, but at the time it was a surprising result. Slough Estates was so encouraged by this report that it became one of the founder members of the Design Build Foundation and the author became its Chairman in 2000. The Foundation had a multidisciplinary membership base and was dedicated to improving D&B performance. It did some ground-breaking work, including involvement with the supply chain initiative, Building Down Barriers (Holti, Nicolini and Smalley, 2000), and was subsequently absorbed into Building Excellence (BE) and Constructing Excellence. The Reading report was the catalyst that enabled D&B to go from a small player to market leader, in procurement market terms, in the space of a few years. SCM played a role

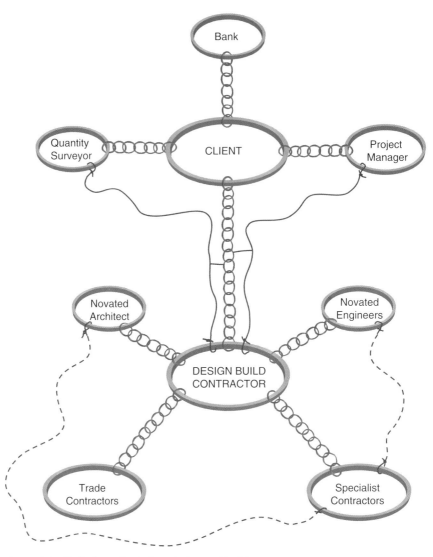

Figure 7.4 Typical supply chain – design and build.

in this success because the supply chain structure is ideal for its exploitation.

7.2.6   Design and Build (D&B)

Figure 7.4 shows a typical supply chain structure for D&B.

This form of procurement provides one of the strongest platforms from which SCM can flourish. A design build contractor is a party to all the important relationships, upstream and downstream, and is in the best position to lead the efforts to add value and reduce costs. Furthermore, when involved in repeat work on similar types of building, design and build contractors can promote continuous improvement and knowledge transfer from one project to the next.

The potential for the successful use of SCM is reduced when clients have a direct relationship with architects, engineers and quantity surveyors during the initial stages of a project, bringing in the contractor later and then novating the designers to that contractor. The later this is done in the process, the more difficult it becomes to introduce SCM successfully. Clients generally seem to believe it necessary to have direct control of design at the start, but are less likely to use novation when contractors demonstrate that design is safe in their hands. Contractors have been ostracized from much upstream activity for so long in the construction industry that it will take some time before they become single-point delivery experts with whom clients and designers are happy to work. Clients need the benefits of SCM, but also need to know that designs are not going to be emasculated by the process. There would be little point in SCM straining to produce the last ounce of waste reduction if the overall design, particularly the aesthetic qualities, suffers as a result, thus greatly reducing value. The Japanese car industry started from the other direction and concentrated on efficiency of production before they realised the importance to the customers of design. Customers deserve both efficient production and good design, and the more sophisticated forms of D&B in the UK can now deliver.

The adoption by Slough Estates of D&B and the way in which SCM ideas were integrated into the process are discussed later in this chapter. In summary, modified versions of D&B proved extremely successful and showed that SCM practices could be used well in the construction industry.

Surprisingly, the Government, despite competition rules and public accountability, has embraced D&B and value-based sourcing (encouraged by Egan) and has extended the requirement of services from contractors to include finance and facilities management. The introduction and widespread use of the Private Finance Initiative (PFI) may have been motivated by the need to reduce public expenditure by spreading costs over a long period, but it coincidentally resulted in a procurement system tailor-made for the application of SCM.

7.2.7   Private finance initiative (PFI)

Figure 7.5 shows a typical supply chain structure for PFI. Gone are any constraints to establishing effective relationships in the supply chain and gone are the design discontinuities. Clients in this system need only establish performance criteria and then the contractors (as part of the Special Purpose Vehicle – SPV) deliver buildings to match those criteria. Contractors have full control of finance and design, and have Facilities Management (FM) expertise within their own team. In the long term, when contractors are able to work on a series of generically similar projects, the conditions are ideal for establishing effective partnering arrangements and systematic continuous improvement. For the first time the construction industry was able come

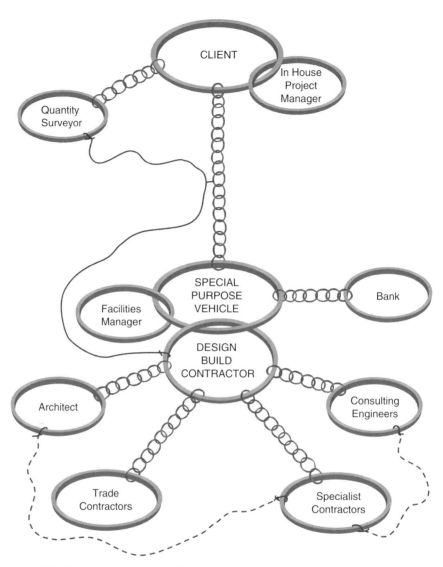

Figure 7.5 Typical supply chain – PFI.

close to matching manufacturing in SCM terms. The arrival of sophisticated D&B and PFI have placed the construction industry in a position where it can perform much better and eventually perhaps even change the poor image it currently enjoys with the public. PFI also facilitates the opportunity to apply through-life costing, avoids the waste of unnecessarily high maintenance costs and deals with sustainability issues in an holistic way.

The choice of procurement route is an important factor in the successful use of SCM in construction projects. In particular, procurement routes which exclude contractors and specialist subcontractors from both strategic and detailed design decisions until relatively late in the overall process of design and construction, make the use of SCM difficult or ineffective.

7.3   Influences of Industry Reports

The two seminal reports commissioned by the Government in the 1990s, the first by Sir Michael Latham (Latham, 1994) and the second by Sir John Egan (Egan, 1998), had together a big impact on the industry and relate to this discussion on the use of SCM in construction.

7.3.1   The Latham report (1994)

Slough Estates was not influenced so much by Latham (1994), since establishing long-term relationships with professionals and contractors, and paying on time, had long been cornerstones of its business philosophy. However, the company had never established formal partnering relationships as proposed by Latham. After much deliberation, it was concluded that there would be no material benefit to formalising the already excellent relationships. Furthermore, it was not possible to commit to providing a certain amount of work to suppliers, since the development programme was so volatile.

The company experimented with individual project partnering charters, but it was difficult to detect if these extra-contractual (in building contract terms) documents had any influence on performance. For other clients who do not have such a strong track record for collaborative working, partnering charters could prove quite useful, provided the way the formal contracts are set up does not make confrontation more likely than collaboration.

7.3.2   The Egan report (1998)

Slough Estates' Chairman, Sir Nigel Mobbs, was a member of the Egan committee and the author was on the support team set up to carry out investigations and produce draft reports. From the beginning, it was clear that Sir John Egan's approach to the problem was totally different from any of his predecessors. It was a revelation to be discussing customer value, product team integration and a quality-driven agenda in a construction forum. It was like having to learn a new language. We now know that the approach and the language were common in the manufacturing industry and Sir John had become a passionate believer in the power of the Total Quality Management (TQM) approach from the way it had transformed the motor car industry. He gave short shrift to any proposition that the construction industry is a special case and produced the sort of report that would not have been possible from anyone inside the construction industry.

The author was involved in visits to organisations, such as Nissan in the North East, that were among the best UK practitioners of TQM and SCM. It was inspiring to witness ordinary people doing extraordinary things, contrasting sharply with some of the finest brains in the country being involved in disastrous results on construction projects. It was also good to hear how inflation cost increases to customers had been replaced by a

Part B

more-for-less product offer each year. Contrast this with the built-in inflation assumptions of a construction cost plan. The immediate lessons that
the author brought back to Slough Estates were:

- Anything whatsoever that does not contribute to customer value (see
  definition earlier) must be regarded as waste, muda in Japanese, and must
  be systematically eliminated.
- In order to maximise value and drive out waste the supply chain must
  be involved in the whole process of design and construction, and the
  operatives who carry out the work within the supply chain should be a
  constant source of information on possibilities for improvements.
- As much as possible of the work on new designs or design improvement
  should take place outside the pressures of a live project, termed 'product
  development'.
- Measuring results, including productivity, and setting targets for improvement creates a platform for becoming best in class. Step change improvements can only be expected from time to time and the accumulative effect
  of many small changes is just as important.

These guiding principles were applied to Slough Estates projects post-Egan
with remarkable results.

7.4   Slough Estates SCM Initiatives and Results

After all the poor experiences with the normal range of construction industry procurement systems and exposure to how manufacturing in the UK had
started to perform much better, it was clear that the important first step that
needed to be taken was to establish a platform from which TQM and SCM
could operate successfully. Slough Estates, with its open-minded attitude,
in-house design and construction resources and strong development programme, was in a strong position to take such a step.

7.4.1   Platform for SCM

It was concluded by the author that the necessary characteristics of a
successful platform would be:

- Consistent strong leadership dedicated to driving the new agenda;
- More open management structures, less command-and-control and less
  bureaucracy, as successfully demonstrated in a social network analysis
  exercise (Pryke, 2001);
- Recognition, respect for and involvement of the people who carry out
  the work on site with systematic feedback on improvement ideas from
  them;
- Investment in product development, measurement of performance and
  sharing knowledge with others in industry networking groups (Design&
  Build Foundation (D&BF), Reading Construction Forum (RCF), Movement for Innovation (M4i), etc);

- Single point responsibility for both main contractors and key specialist contractors;
- Appropriate commercial terms for each relationship with the emphasis on open-ness, collaboration and negotiation.

The author was so convinced that there was huge hidden waste in the normal construction processes that he chose to spend a high proportion of his time driving the necessary changes and encouraging others to join in the efforts. However, one of his main roles also was to keep the Chairman and the Board informed about progress in an area in which some of them felt uncomfortable, particularly on matters like abandoning competitive tendering for main contracts. Some were not convinced until the benchmarking results later started to show how beneficial sensibly applied financial negotiation coupled with highly effective design management can be. The tendency for UK clients to trust professionals and mistrust contractors is strong and presents one of the main obstacles to be overcome by those wishing to run construction projects along collaborative lines.

7.4.2   Supply chain structures

Projects were carried out at Slough Estates in the 1990s in two ways. If projects suited the company's internal resources in architectural design, they would be structured as Figure 7.7. Otherwise projects would be structured as Figure 7.6. In both arrangements, the projects were allocated a development manager, who generally came from a general practice Chartered Surveyor background, and an internal project manager who was either a Chartered Engineer or Chartered Quantity Surveyor. Engineering resources were always procured externally. Projects were nearly always funded from the company's own resources and nearly always built as additions to the portfolio for rent. There were occasional projects built to trade on, but these were handled by separate management in the Slough Estates group. It is interesting to note, in this respect, that the institutional funds providing finance for development projects would generally only work with traditional forms of procurement. As a result, projects developed by the department in Slough Estates that built for sale (and needed external sources of finance to build) were not able to exploit SCM as effectively as the department which built for rent (and funded projects from internal resources).

The in-house design and construction supply chain (see Fig 7.7) shows clearly how direct relationships were established with all the major suppliers, including labour-only contractors. This enabled direct dialogue to take place between client and workforce – inconceivable (and contractually undesirable perhaps) in the normal procurement systems. The management of the relationships was able to be driven by Slough, just as it is by manufacturing clients, and there were few impediments to introducing SCM tools and techniques that were felt appropriate. This situation was probably unique in construction and it is fortunate that senior management at Slough

Part B

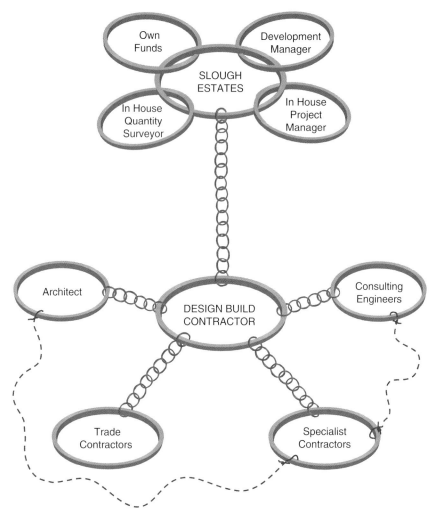

Figure 7.6 Slough Estates supply chain – external projects.

encouraged experimentation to show what is possible in the industry in a truly integrated environment.

7.4.3   Management of relationships

It is worth repeating at this stage the SCM definition used for this chapter.

> *The management of upstream and downstream relationships with clients and suppliers to achieve greater project value at less cost.*

At Slough Estates, the author set out the strategy for how each project would be organised, decided the sort of commercial arrangements that would be

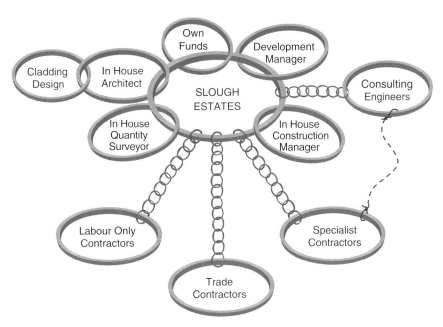

Figure 7.7   Slough Estates supply chain – in house design and construction.

entered into and took part in the selection of, and negotiation with, key project actors. The project manager managed the project based on the strategic guidelines but the author was constantly involved in the management of the relationships, establishing strong contacts at Partner or CEO/Director level to make sure suppliers understood the new ways of working and would drive the cause from their end. Internally, as discussed earlier, the author made sure senior management and the 'internal' client, the development manager, were kept informed about progress in the new ways of working. It cannot be stressed too strongly that both internally and externally people were being asked to perform outside of their comfort zones, and it was necessary to maintain constant enthusiasm in order to keep doubters from bringing damaging negativity to the process.

7.4.4   Commercial principles and contracts

The underlying commercial aim was to harness the best resources in design and construction in such a way that the team worked together collaboratively at improving client value and reducing costs in the knowledge that their margins were relatively secure. The principal contractual aim was to make main contractors on external projects and specialist contractors on all projects completely responsible for product delivery and performance, with no split responsibilities for design. These aims were successfully achieved as follows:

- Single point delivery main contracts and contracts with specialist contractors were all negotiated following interviews and the provision of information on overhead and profit expectations by the contractor.
- On external projects, architects and engineers chosen by Slough were engaged by the contractor from the beginning. Until a contract was signed, their initial payments were covered by Slough and paid through the contractor on a cost reimbursable basis. Any such sums were then included in the negotiated figure with the contractor for fees. This procedure avoided the use of novation – a costly technique that is often used as a risk-shedding device rather than a genuine attempt at integration of the process.
- Where a specialisation, such as M&E contracting, benefited from consulting engineering input, the consulting engineers were integrated into the M&E contractors' team. This allowed the M&E specialist to take full responsibility for the installations and their performance, with the client paying for the consultants' input.
- The cost plans comprised agreed overheads and profit and target costs for each part of the work. The team was then given the task of designing and planning the work to lower these costs as much as possible without undermining the product quality and value. The final lump sum agreed was often left until late in the project to maximise the potential of this process.
- Any suppliers that had no design input supplied goods or services in direct price competition or as part of a term agreement to supply all projects over a certain time scale.
- Contracts were kept as simple as possible (typically one or two pages of A4 only), with suppliers to internal projects eventually being employed on the basis of a purchase order.
- There were no direct financial incentives applied in any part of the process. The healthy incentive for suppliers was the prospect of obtaining repeat business in a financially secure environment. Slough was known as the best paymaster.
- Where contractor or specialist CEOs committed personally to resolving any residual defects, the contracts with them did not have a retention clause. This gave emphasis to the aspiration of 'right first time' and was in the spirit of a zero defects philosophy.

This commercial framework encouraged suppliers to take part positively in the new ways and a new spirit was engendered. This was quite different from the defensive, aggressive and secretive postures seen in normal commercial life in construction.

7.4.5   Achieving greater project value

For the purposes of this chapter value is defined as:

Any characteristic, feature or performance (of a design, building or project) that is important to the customer.

It is design that creates much of the value in construction, and good design responds to predetermined values of the customer (the brief).

Slough Estates has always invested considerable time and effort in understanding and responding to the needs of tenants. Its hands-on internal property management team continually fed back customer responses to the buildings they were renting. In addition, the development team and internal architects were always seeking improvements ahead of the market and pioneered design in flexibility/adaptability, common service pods, and glass dividing walls in industrial units. Tenant value was thus well catered for and the additional input in this area from designers was limited. The other part of the value equation was the additional value created for Slough Estates as an investor, such as the durability of the product to cater for demand beyond the first tenants. Again the internal team had developed a strong set of criteria in this area, particularly with respect to architectural design durability, as a result of which the company had a distinctive brand for building design. Therefore the management of the design supply chain was mostly downstream with the focus on delivering to these established tenant and investor values.

The benefit that Egan brought to value management was to concentrate everyone's mind on what was really important to customers rather than what was assumed to be important. One of the great weaknesses of the construction industry is that too many assumptions are made on behalf of customers and there is a great deal of over-design and over-specification as a result.

7.4.6   Product development

Another of the Egan proposals was that development of designs and products is best carried out independently of live projects. With its internal team and financial strength, Slough was in a position to experiment with product development and the author chose to set up a team to work on industrial roofing and cladding. This was a good field to start with because the company's main cladding partner/contractor had ceased trading (another victim of the industry's normal aggressive commercial environment) and Slough had taken the opportunity of adding its chief designer to the internal architects' team. Also, cladding materials were being purchased direct and the fixers of the old company were employed on a labour-only basis. Every resource was under the direct control of Slough and a team was set up to consider how the current system and products were working and how best they might be improved. The team consisted of representatives from the internal design and construction team, the consulting structural engineer, profiled metal manufacturers and the cladding fixers, now self-employed, and was truly representative of all parties with an interest in cladding.

Product development meetings were scheduled once a month after an early breakfast. After a hesitant start for the fixers, they soon came to dominate the dialogue. It quickly became apparent that they knew more about the process than anyone else and they were eventually happy to let

Part B

others share their knowledge. It was clear there was much inefficiency, mainly caused by design details and material scheduling. Some details were proving almost impossible to construct and materials were delivered in an order that wasted unpacking time. This 'muda' would not have come to light without such an opportunity for dialogue.

The main results from these meetings were:

- Cladding design details were changed to improve buildability and safety.
- Material scheduling and delivery were designed to suit production.
- Initial production information was produced much earlier in the design process.
- The fixing bought into the zero defects philosophy, agreeing that inspection by third parties was wasteful and unnecessary.

The same team was instrumental in Slough becoming one of the first users of terracotta tiles on commercial buildings in the UK. They developed an inexpensive support system for the tiles and an innovative fascia system for application on a flexible-use building. The power of a multi-disciplinary team working together was nothing short of miraculous in construction terms, and the bonus was a dramatic improvement in the quality of the site personnel's working lives. The innovation also added value for both tenants and client through increased flexibility in the use of the space created by the development.

On external projects the best example of product development was in air-conditioning installations in business park units, albeit the development took place during real projects over a number of buildings. The coming together of the M&E contractor with the consulting engineer as partners enabled a 'technology cluster' (Gray, 1996) to be formed with the contractor in the lead role. They were given the task of designing installations, the bulk of the jointing and fabrication for which could be carried out in a factory environment. The team designed pods containing pipe-work, fan-coil units and controls that could be fixed rapidly to the ceilings and joined with semi-skilled labour. This facilitated shorter, more reliable installation periods and better quality, and was a fine example of collaborative working. Furthermore, the challenges that were made in open forum to the value being created by each design or component of the system resulted in the removal of over-specified items and contributed to an overall saving of 30%.

These examples can be considered as contributing to 'lean' design (Ballard and Zabelle, 2000) since they avoid the wasteful redesign processes that characterise the normal design-tender-construct processes.

Production methods re-engineering

As part of the post-Egan 'waste watching' initiative, some of the obviously wasteful methods of construction were examined. One of these was the use of scaffolding for cladding and curtain walling industrial units. It was observed that industrial projects begin quickly and the steel frames are erected in a few days, but once the scaffolding is erected the projects seem to slow down perceptibly. Furthermore, scaffolding erection and dismantling are hazardous activities, and the sites are difficult to keep tidy when scaffolding is in place. Mechanical handling equipment was investigated but was initially rejected by the internal estimators/QSs as being too expensive. However, in the spirit of making progress and the willingness to invest in innovation, it was decided to go ahead and the instruction was given that no scaffolding could be used on the next projects except in constricted areas around boundaries. In short, this change improved every aspect of production and ultimately cost less. The sites had to be made level and 'stoned' (surfaced for vehicular access) early in the process to enable machines to operate; this rendered sites safer and more efficient places on which to work. Specialists, such as the window/curtain walling contractors, also benefited from the change, despite their initial reluctance to dispense with scaffolding. This exercise shows that innovation in construction is often difficult to implement unless risks are taken with costs. In the normal project deliberations, if a new idea cannot be costed with any degree of accuracy it is unlikely to be adopted. The process innovation required changes to construction practices, an intelligent reappraisal of the ways in which both costs and value are calculated and negotiations with access equipment providers to ensure that continuity of access could be provided in a manner that overcame occasional equipment failure. The implementation of the changes required information and knowledge sharing across the supply chain in a manner that is rarely found in construction projects.

7.4.7   Measuring waste and benchmarking costs

During this innovative period the Building Research Establishment (BRE) launched a productivity sampling service for construction sites called CALIBRE and found that the number of productive hours spent on value creating activities was typically 35% of the total hours spent by operatives. This was a shocking statistic and Slough Estates immediately employed BRE to measure productivity on two of its adjacent sites. The results fortunately showed that collaborative working and better construction methods produced about 55% of value producing hours. This still seemed low and the feedback obtained from the BRE measuring process, where the sampler investigates instances of low productivity, showed why. Much of it was due to design errors, and involved reworking, defect rectification and waiting for materials. This information, not normally available to senior management, was fed back daily and promoted changes to systems and methods that started to push the value hours higher. One of the techniques that construction in general has taken on, the Last Planner (Ballard and Zabelle, 2000), has helped considerably in reducing lost hours waiting for materials.

In order to monitor the effects of the various initiatives, Slough Estates employed an independent consultant QS to benchmark the costs of the industrial units that it was building against the cost of units being built by competitors – the majority of which were still being procured with separate design teams and a competitively tendered main contract. The results showed that units designed and constructed internally were being built for approximately 20% less than the general market. Furthermore, like for like, the Slough products were perceived (by the independent consultants) to be of a higher specification. It is a source of regret, however, that more was not done in this important benchmarking area so that results could have been published.

7.5   Summary and Conclusions

The normal procurement environment and divisive culture in UK construction has until recently made the application of techniques such as Total Quality Management and Supply Chain Management difficult to apply. In particular, the separation of design from construction and the insistence on competitive tendering has made the collaboration necessary to use these tools impossible to achieve. However, following the Egan report and the Government decision to operate value-based procurement, opportunities are now available in processes like PFI and negotiated design and build for the exploitation of techniques such as SCM.

At Slough Estates in the 1990s a fertile environment existed for experimentation in techniques for improving value and driving down costs. The company had the confidence, resources and desire to contemplate better ways following its own bad experiences with standard industry procurement practices. Slough accepted that integration of design and construction resources, combined with collaborative working, was the starting point. This demanded negotiation with key players to make it most effective and the Board, led by a Chairman determined to see progress in this area, accepted that competitive tendering need not apply, provided that the results were carefully monitored and reported back.

The Slough Estates' in-house design and construction team had direct relationships with key suppliers and was able to use a number of the techniques suggested by Egan, including product development, waste measurement and off-site component production. In total, the initiatives helped the company to reduce costs by about 20% compared to market comparisons. It showed beyond doubt that major improvements are possible in construction through the use of SCM, but that the types of processes which need to be adopted require extraordinary client leadership and commitment to supply relationships that appear on face value to be non-competitive. However, changes in leadership of client businesses, such as that which occurred at Slough Estates in late 2006, are bound to make continuity of such commitments difficult to maintain and much of the promise created by the progress at Slough in the 1990s could be lost. The new Slough Estates' management policy is to outsource as much as possible of non-core property resources, and the in-house design and construction team has been disbanded. This is sad, but understandable, and illustrates the fact that the majority of clients are likely to be more comfortable with distancing themselves from deep involvement in the process, despite the potential commercial advantages of the magnitude obtained by Slough Estates in the 1990s.

The inevitable conclusion of the foregoing is that the construction industry is unlikely to be able to transform itself across the board by using SCM-type techniques, unless sufficient private clients are persuaded to provide the leadership such as that provided by Slough Estates in the 1990s. Furthermore, Government and public sector clients need to keep their nerve and let SCM-friendly processes like PFI mature into the world class deliverers of projects they are capable of becoming.

References

Ballard, G. (2000) *Last Planner System of Production Control*. PhD Thesis, University of Birmingham, Birmingham, UK.

Ballard, G. and Zabelle, T. (2000) *Lean Design: Process, Tools, and Techniques*. White Paper 10, Lean Construction Institute, Oakland, CA, USA.

Bennett, J., Pothcary, E. and Robinson, G. (1996) *Designing and Building a World Class Industry*. Centre for Strategic Studies in Construction, University of Reading, Reading, UK.

British Property Federation (1983) *Manual of the BPF System for Building Design and Construction*. British Property Federation, London, UK.

Cassel, M. (1991) *Long Lease: The Story of Sough Estates 1920–1991*. A Pencorp Book, London, UK (Available directly from Slough Estates).

Christopher, M. (1992) *Supply Chain Management: Strategies for Reducing Costs and Improving Service*. Pitman Publishing, London, UK.

[Egan report] DETR (1998) *Rethinking Construction: The Report of the Construction Task Force to the Deputy Prime Minister, John Prescott, on the scope for improving the quality and efficiency of UK construction*. DETR, London, UK.

Gray, C. (1996) *Value for Money* Reading Construction Forum. University of Reading, Reading, UK.

Holti, R., Nicolini, D. and Smalley, M. (2000) *Building Down Barriers: The Handbook of Supply Chain Management-The Essentials* .CIRIA, London, UK.

Latham, Sir M., (1994) *Constructing the Team: Joint Review of Procurement and Contractual Arrangements in the United Kingdom Construction Industry*. HMSO, London, UK.

O'Brien, W. (2001) *Construction Supply-Chain Management: A Vision for Advanced Coordination, Costing, and Control*. University of Florida, Gainesville, USA.

Pryke, S.D. (2001) *UK construction in transition: developing a social network approach to the evaluation of new procurement and management routes*. PhD thesis Bartlett School of Graduate Studies, University College London.

Vrijhoef, R. and Koskela, L. (1999) *Roles of Supply Chain Management in Construction*. University of California, Berkeley, CA, USA.

From Heathrow Express to Heathrow Terminal 5: BAA's Development of Supply Chain Management

Keith Potts

This chapter covers the significant contribution that British Airports Authority (BAA) has made to the development and exploitation of supply chain management (SCM) philosophies and techniques. In the limited space available, the chapter deals with a fifteen year period during which BAA committed considerable resources to the development of its SCM strategies, culminating in the completion of Heathrow Terminal 5 shortly before this book was published.

In 1996 Sir John Egan, BAA's Chairman at the time, was looking for BAA to identify and embrace *World Class Procurement* in order that this could be fully implemented on the Heathrow Terminal 5 project. This £4.3bn five-year mega-project, with a projected finish date of March 2008, was to be one of the largest construction projects in Europe and the largest project ever attempted by BAA.

Indeed as the cost of construction of T5 was the largest single threat to BAA's survival, BAA decided to fundamentally rethink the construction process in order to avoid the cost and time overruns which were the norm for mega-projects in the UK (Building Magazine T5 Supplement, 2004). Typical examples of mega/large projects in the UK, which all finished late and over budget included the British Library, upgrade of the West Coast Main Rail Line, London Underground's Jubilee Line Extension, the Scottish Parliament building and the new Wembley Stadium.

BAA investigated the causes of failure in mega-projects and found that the cause in many cases was the way in which the supply chain was engaged and the way risk was managed. BAA considered that the traditional UK construction industry approach of transferring risk to the supply chain would not work on the T5 project. A new approach, requiring the client (BAA) to perform a dominant central role within the procurement and management process, was implemented. This process approximated to a construction management approach with BAA acting as the lead construction manager in collaboration with a core team of consultants and contractors.

The T5 project management philosophy did not occur overnight but was the culmination of over a decade of continuous trials, reviews, reflections and incremental improvements by the BAA project team. Critically, during the first decade of the new millennium, BAA had developed strong links with a regular team of framework partners who worked together with their client from project to project to drive down the cost and increase the quality and the efficiency of their work (New Civil Engineer T5 Supplement, 2004).

8.2   Heathrow Express High Speed Rail Link

The Heathrow Express is a privately operated high-speed service link offering the fastest journey time between Heathrow Airport and Paddington Station in central London – 21 minutes to Terminal 4. The £350m project originated in 1993 as a joint venture between BAA and British Railways Board, designed to increase use of public transport to and from the airport from 34% (already the highest in the world) to 50%.

Balfour Beatty was appointed as main contractor on one of the first UK major projects based on the then recently published NEC Conditions of Contract. The payment approach was based on a lump sum contract based on Option A – *Priced contract with activity schedule*. This innovative form of contract, which was endorsed by Sir Michael Latham in his Report *Constructing the Team*, was designed to encourage a spirit of cooperation and team working.

In October 1994 a major tunnel collapse occurred in the Central Terminal area of the airport. The collapse caused a huge crater to appear between the airport's two main runways and caused damage to car parks and buildings. The contractor Balfour Beatty was using the New Austrian Tunnelling Method (NATM), which involved spraying fast-setting concrete onto steel mesh reinforcement to support the tunnel. This technique had been used extensively throughout the world on major tunneling projects. However there was relatively little experience of working with NATM in the England – particularly in London clay (Sauer, 1994).

The collapse at Heathrow was a potential disaster for BAA, which plunged the entire project into crisis. Against a background of cash flow anxieties, a worried supply chain of suppliers and subcontractors, a hostile media, safety issues and threats of litigation, a recovery plan was worked out. BAA resisted the opportunity to sue the contractor and instead adopted an informal partnering approach, which reinvigorated the entire project team and workforce (MPA Project Recovery seminar, June 2000).

A *solutions team* was set up to review working methods comprising representatives of the client, the contractor, the designer and the project insurers. The aim was to work together in a co-operative framework within a *no blame* culture in order to minimize delay. The parties became known as

Part B

the single team using the best people for each task, regardless of the parent company affiliations.

Critically, following the tunnel collapse, the decision was taken to change the payment scheme to Option E *Cost Reimbursable* with a side agreement between the parties to share cost over-runs (Finch and Patterson, 2003).

BAA appointed a new Construction Director for the recovery project who demonstrated and promoted a culture of cooperation instead of rivalry. He employed '*a team of specialist change facilitators and behavior coaches to work with the team on changing the tradition of adversarial working in order to make the single team 'statement of intent' a reality. This was demonstrated in the everyday behavior of the construction team*' (Lownds, 1998).

At one point, the Heathrow Express project was 24 months behind schedule, but eventually became operational only nine months after the original project completion date. The lessons learned from the Heathrow Express Tunnel collapse were significant and led to an extension of the concept of cooperation within the NEC contract to informal partnering based on an integrated team approach. Overcoming the disaster on the Heathrow Express project was considered a huge success within BAA (Brady *et al.*, 2006).

Balfour Beatty pleaded guilty to failing to ensure the safety of both its employees and members of the public and was fined a record £1.2 m. Austrian engineering firm Geoconsult, which was responsible for monitoring settlement on the Heathrow Express Link, was also fined £500,000. Both companies were ordered to pay a further £100,000 each in costs. Justice Peter Cresswell described the incident as '*one of the worst civil engineering disasters in the UK for last quarter of a century*' (BBC News, Monday 15 February 1999).

8.3   Continuous Improvement of the Project Process (CIPP)

In September 1995, BAA introduced a standard set of guidelines summarized in its *Project Handbook*. The guidelines were intended to improve the project development and project management process by ensuring 'a consistent approach to projects across the group which meets business needs and opportunities through the optimum business solutions'. This improvement exercise was referred to as *Continuous Improvement of the Project Process* or CIPP. The CIPP handbook aimed to establish a consistent best practice process and applied to all projects with a value over £250,000.

> '*The CIPP handbook laid down a set of key policies or principles – these deal with safe projects, a consistent process, design standards, standard components, framework agreements, concurrent engineering and pre-planning. All capital projects had to adhere to these policies. The handbook provided a template for the organization of BAA projects, and outlined a seven-stage process covering the project life cycle from inception through to operation and maintenance. Each stage included a series of checkpoints, which had to be completed, and a series of evaluation gateways where the process was assessed by an evaluation team*

before going to approval gateways for sign-off from local and/or group capital project committees. To successfully pass through a gateway, 8 key sub-processes needed to be managed and the outputs from each co-ordinated: development management, evaluation and approval, design management, cost management, procurement management, health and safety, implementation and control, commission and handover. BAA developed a process map showing each stage along the top and the sub-processes within each stage, outputs and gateways (Brady et al., 2006).

Postscript: In 1999 BAA took a hard look at the CIPP process and recognized that the process still reflected traditional construction practices. What was needed was a process, which reflected the shift from construction to assembly and the need to design for manufacture. This approach reflected the move away from one-off bespoke construction towards a more predictable design, manufacturing and assembly process (McLeod, 1999).

8.4   SCM at BAA: 'The Genesis Project'

BAA's Genesis project provided the test bed for a new project process governed by three imperatives: the need to reduce costs, shorten delivery times and improve quality. A relatively simple project with a robust business case was selected as the test bed project – a five-storey thousand vehicle car park building with offices and accommodation for the World Cargocentre. The £8m project was let based on the NEC Engineering and Construction Contract and was completed in March 1997, after a nine month construction period. Each of the packages was let on a *with design* basis with different terms and roles for the participants compared to traditional procurement (Pryke, 2001). The level of involvement of the client was reflected in the overheads allocated to the project by BAA – in the region of 20%, for what was essentially a design and build project (Pryke, 2001).

Developing a World Class Project Process required a fundamental change in the role of the project team, procurement methods, workflows and communication systems. The project team established itself as a fully open and integrated team (including downstream suppliers), able to work on all aspects of the design concurrently. By developing a seamless interface between the client and the team, knowledge transfer became more fluid, enabling learning within a *virtual company* environment. The project team developed a mission statement, team values and a project charter in order to generate ownership and accountability and ensure that the project aims and objectives remained to the fore (Mace, 1997).

The new process aimed to achieve an ambitious range of objectives including the development of an integrated team approach, mapping the supply chain, developing the supply chain management, introducing component based design and developing the process for productivity improvement (BAA, 1997). Figure 8.1 shows the contractual and management framework with the expert client BAA at the centre of the hub.

Part B

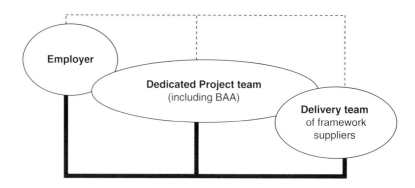

BAA contracts directly with all suppliers

---------- Management by most appropriate member of project/delivery team

Figure 8.1 BAA contractual and management arrangement: now and in the future. Source: BAA Frameworks: Project guideline adapted from BAA (1997), cited in Pryke (2001).

BAA realised that it would have to take a radically different approach with the creation of a self-management environment with accountability right at the point of decision-making. This would require passing decision-making right down the line and empowerment of not just the line management team but also the supply chain. '*An example of this was that the brief was not handed down from on high, but was written and therefore owned by the team*' (Williams from E.C.Harris cited in BAA 1997). Under this arrangement, subcontractors were elevated from a position of *receiver of instructions to collaborators in design and planning decisions* (Pryke, 2001).

BAA spent a long time mapping out processes, before carrying out the tasks, and also identified everyone's strengths and weaknesses, before identifying the roles and responsibilities, noting the gaps and duplication of effort. BAA also identified the major elements, which contributed to the bulk of the cost and looked at the key drivers – on this project the structural frame and the mechanical and electrical services installations (M and E). The supply chain maps, which were developed with the key suppliers, enabled real opportunities for improvements to be identified.

Each tier of the supply chain was analysed to assess the opportunities to add value, and classified into *High*, *Medium* and *Low*. Those maps, which were developed in most detail, proved the most useful and offered the greatest opportunity for savings to be reviewed in detail. For example, the O'Rourke map for the *Substructure* included an activity cost breakdown into *Labour, Plant, Materials and Sundry*.

BAA believed that the historical roles and traditional plans of work and standard traditions of engagement would have to be changed if 30–50% savings were to be secured. They considered that the traditional approach with one person responsible for managing the whole project as inappropriate, believing that the project should be separated into manage-able areas demanding different levels of expertise.

The Genesis project was thus split into substructure, infrastructure, car park, core and retail areas with BAA identifying the team leader and the team members within the work section. Under this arrangement *delivery teams* were formed at the Concept Stage to manage execution of the design and construction for a particular zone of the project. The delivery team took ownership and took the project right through design to completion and were accountable for their cost plans, their programme, their safety target, their logistics and productivity. There was also a *coordination team* to liaise between the delivery teams. These delivery teams were an early application of the concept of *work* or *technology clusters* in construction (see Gray, 1996; Holti, 1996; Holti *et al.*, 1998)

Fundamental to the success of the process was the earliest possible involve-ment of the supply chain members in the design process. By introducing these specialists at the beginning of the concept stage, the production-ori-entated knowledge of construction contributed to the design process from the outset. The whole team worked together under the *Genesis* name with no reference to each company's identity. The team shared resources includ-ing computer systems and operated together from a central office on site.

The Genesis project set out to build using a database of standard com-ponents with a maximum use of modularised and preassembled off-site components. This approach also had benefits for the safety programme, the productivity and the attitude of the workforce. Sourcing the suppliers early in the concept stage and developing work breakdown structures in detail to form supply chain maps secured significant savings. '*This mapping was one of the key successes in the project and it even went as far as the second and third tier*' (Reynolds from MACE, cited in BAA 1997).

The original cost plan and budget for the project increased from £7,593,000 to a project final cost of £8,229,579. The cost increases were related to three primary factors: additional service and highway diversions; the escalation of design consultants' on-costs and prolongation costs as a result of programme delays. The design consultants' tasks and duties were not clearly defined and did not match the needs of the project (Pryke, 2001).

8.5   BAA Initiatives

In 1997 BAA received a special award at the British Construction Awards for being the '*most successful client in the construction industry in the last ten years*'.

The BAA Construction Report 1997/98, stated ' . . . *we want to share our experience with all who purchase or supply construction services*'. The Report identified three key issues: *partnering*, *innovation* and *best practice*.

Partnering: One of the first partnering initiatives in BAA was with the Pavement Team, which builds or restores runways, taxiways and airport stands (New Civil Engineer, 1997). The selection process began in Autumn 1994, following Sir Michael Latham's call (Latham, 1994) to the industry to cut its costs by 30%. The selection process took over a year and Amec Civil Engineering was eventually chosen under a five-year serial partnering arrangement. Prior to this arrangement BAA had experienced a number of problems with the construction and maintenance of runway pavements. High costs were often a result of generally adversarial and fragmented procurement adopted by the airport operator and its contractors (Cox and Townsend, 1998).

Under the new partnering approach BAA were required to work closely with the best contractor and suppliers in an integrated team taking a proactive role in the management of the supply chain. The relationship took on a separate identity to either BAA or Amec, becoming known as the *Pavement Team*. The Pavement Team was, therefore, an integrated project organisation involving staff from both client and contactor's organisations.

The form of contract governing the relationship between the client and contractor within the Pavement Framework was the New Engineering Contract ECC Option C – *Target Cost with Activity Schedule*. This approach allowed for target costs to be established, and variations to be dealt with as compensation events using open-book evaluation (Cox and Townsend, 1998).

The pavement projects were considered more like a manufacturing process than a traditional construction approach. The project team carried out *process mapping* under which processes were reduced to a simplistic flow chart of activities, with each activity and sub-activity carefully analysed for productivity, and non-productive standing time with the object of designing out the inefficiencies. Further benefits of the partnering approach included the necessity to keep one set of records and a joint quality control system, improved conditions for site workers and higher levels of safety. The long-term partnering approach also encouraged innovation.

Analysis of the Pavement Team's projects showed a significant improvement in cost predictability, compared to similar contracts carried out before the partnering agreement.

Innovation: One of the successful outcomes of the Genesis project was the development of a productivity tool kit (Eke cited in BAA 1997). In conjunction with the Building Research Establishment (BRE), the Genesis project was used to trial a process aimed at identifying productive and non-productive working by mapping all the site processes and identifying and coding all the tasks. Each operative was monitored on a regular basis by an independent observer from BRE and the results logged against a coded structure.

This innovative performance monitoring software system, called CALIBRE (see www.bre.co.uk/page.jsp?id=360), enabled information to be generated on a daily, weekly or monthly basis. Unlike conventional work study exercises the CALIBRE system was developed to study the whole site process, rather than individual tasks in isolation. The user-friendly reports enabled the management team to see at a glance what was happening on the site. Items marked *red* indicated non-productive working, *blue* indicated productive working, *light blue* indicated productive support time (such as setting out) and *green*, statutory breaks. This information enabled the management team to identify how much time was spent on activities that directly add value to the construction and how much time was being wasted on non-productive activities. Having captured the data, this information could then be used as a benchmark for future projects.

8.6   Best Practice – Framework Agreements

In 1993–94 BAA began its framework programme to partner with a number of preferred suppliers on an ongoing basis. The five-year agreements provided suppliers with the opportunities to learn and they included incentivised performance targets, which challenged the suppliers to make continuous year-on-year improvements. The framework agreements embraced a wide range of services including design and engineering consultancy, construction and specialist services.

In 1998, BAA recruited Tony Douglas as the Group Supply Chain Director. At the time of his appointment, BAA had 26,000 suppliers with 23 different processes and 17 different systems for managing the transaction. With 24 different architects, 23 cost consultants, 70 or so external project managers and 340 construction suppliers, slimming down the supply chain was one of his first challenges.

In 2002, BAA developed a second generation of Framework Agreements to achieve more accurate project cost, to implement best practice, and to work with suppliers in longer-term relationships. These Framework Agreements sourced the *best-in-class* capability and were valid for a 10-year period. Suppliers worked with BAA in integrated project teams to cultivate close co-operation, to leverage the right expertise needed for specific projects and to reduce costs (Brady *et al.*, 2006). Critical to the second generation Framework Agreement was demonstration of hard evidence of commitment to continuous improvement.

8.7   Motivations and Influences

8.7.1   The Machine that Changed the World

One of the architects of the *World Class Procurement* approach piloted on the Genesis project revealed that the team were influenced by the work of Prof. Dan Jones (see *The Machine that Changed the World* by Womack, Jones and Roos, first published in 1990 and *Lean Thinking* by Womack and Jones, first published in 1996).

The ground-breaking book *The Machine that Changed the World*, reviewed on a global basis the performance of car manufacturers automakers operating under the traditional mass production basis compared with those operating on the (then) revolutionary lean production system. Their in-depth analysis identified that the Japanese Toyota plant was almost twice as productive and three times as accurate as the US General Motors plant using 40% less manufacturing space and with defects reduced by a factor of three. The key factors in implementing a lean production system were identified (Brady *et al.*, 2006):

- Utilisation of dynamic multi-skilled work teams containing all the relevant expertise with strong respected leaders (*shusa*);
- Creative challenge of constant improvement and problem solving with all workers seeking to add value with a vast reduction in the numbers of indirect workers;
- Implementation of *the five whys* method of problem solving (every error systematically quickly traced back to its ultimate cause;
- All relevant production information displayed on *andon* boards (lighted electronic displays);
- Selecting suppliers on the basis of past relationships and proven record of performance, rather than lowest bid;
- Basic contract, which lays down the basis for a cooperative relationship; parties want to work together for mutual benefit;
- Use of target cost approach with the application of value engineering and value analysis techniques with suppliers keeping all the profits derived from its own cost saving achieved by incremental improvements (kaizen) or innovations;
- Suppliers organised into functional tiers with first tier suppliers producing an integral part based on a performance specification;
- Sharing of knowledge with other members e.g. between first team suppliers and with assemblers' personnel working with suppliers;
- Willingness of suppliers to come up with innovations and cost saving suggestions and work collaboratively;
- Long-term relationships built on mutual interdependence and cooperation based on a rational framework for analysing costs, establishing prices and sharing profits;
- Utilising the just-in-time (*kanban – the means through which JIT is achieved*) system for the flow of parts and production smoothing (*heijunka*). This approach has the additional benefits of fewer inventories and less manufacturing space being required;
- Reduction in waste (*muda*) and cutting costs through the use of IT;.
- Focus on the customer – exceed their expectation.

In the mid 1990s, senior BAA managers visited the Lean Construction Institute in Stanford, California and spent time with leading clients from other sectors such as Tesco and MacDonalds from retailing, and Nissan, Rover and Unipart from the car industry (Brady *et al.*, 2006).

8.7.2   The Egan Report 'Rethinking Construction'

Sir John Egan, BAA's Chief Executive from 1990 to 1999, was instrumental in transforming BAA's project performance. Prior to joining BAA Egan had been the Chairman and Chief Executive of Jaguar Cars and had witnessed at first hand the radical improvements which could be made in the automobile industry when the principles of lean construction were embraced. Indeed, during his time with Jaguar Cars Egan had successfully turned the company from virtual bankruptcy into a £1.6bn prize for Ford (http://www.locum-destination – cited 18 September 2007).

In 1997 Sir John Egan was invited by the Deputy Prime Minister to chair the Government's Construction Taskforce, which was charged with identifying the potential for introducing improved working practices throughout the industry. The key recommendations within the report *Rethinking Construction* (DETR, 1998), which became known as the Egan Report, challenged the industry and its major customers to rethink construction so as to match the performance of the best consumer-led manufacturing and service industries. Bennett and Baird (2001) summarized the key issues involved in the Egan Report as follows.

- Integrated processes and teams should be introduced as a key driver for change.
- The industry should organise its works so that it offers customers brand- named products, which they can trust to provide reliably good value.
- The industry should work through long-term relationships using partnering, which aims at continuous improvements in performance.
- Benefits from improved performance should be shared in an openly fair basis so that everyone has real motivation to search for better answers.
- Project teams should include design, manufacturing and construction skills from day one so that all aspects of the processes are properly considered.
- Decisions should be guided by feedback from the experience of completed projects so that the industry is able to produce new answers that provide even better value for the customer.
- Standard products should be used in designs wherever possible because they are cheaper, and in the hands of talented designers, can provide buildings that are aesthetically exciting.
- Continuous improvements in performance should be driven by measured targets, because they are more effective than using competitive tenders.
- The industry should end its reliance on formal conditions of contract, because in soundly based relationships in which the parties recognize the mutual interdependence, contracts add significantly to the cost of projects and add no value to the customer.

Under Egan's guidance, BAA's senior managers began applying the principles laid out in the Egan Report to improve BAA's project processes and relationships with suppliers and by the late 1990s BAA had made significant improvements to its project execution capabilities, reflected in a greater degree of predictability in terms of time, cost and quality (Brady *et al.*, 2006).

Part B

8.8   SCM on Heathrow T5

The BAA Heathrow Terminal 5 was one of Europe's largest and most complex construction projects. The Secretary of State approved terminal 5 on 20 November 2001 after the longest public inquiry in British history (46 months). The five-and-a-half year site programme commenced in December 2002 and when completed in March 2008 T5 added 50% to the capacity of Heathrow.

The £4.3bn project included not only a vast new terminal and satellite building but nine new tunnels, two river diversions and a spur road connecting to the M25; it was a multidisciplinary project embracing civil, mechanical, electrical systems, communications and technology contractors with a peak monthly spend over £80 million employing up to 8,000 workers on site. The construction of T5 consisted of 16 main projects divided into 140 sub-projects and 1,500 work packages on a 260 hectare site.

Phase 1 construction of Terminal 5 was programmed for five years and can be broken down into five key stages (BAA T5 Fact Sheet):

- Site preparation and enabling works (July 02–July 03) – preparing the site for major construction activity. The work included a significant amount of archaeological excavation, services diversions, levelling the site, removing sludge lagoons and constructing site roads, offices and logistics centres.
- Groundworks (Nov 02–Feb 05) – included the main earthworks, terminal basements, connecting substructures and drainage and rail tunnels.
- Major structures (Nov 03–Sept 06) – the main terminal building (concourse A), first satellite (concourse B), multi-storey car park and ancillary structures.
- Fit out (Feb 05–Sept 07) – significant items of fit out included building services, the baggage system, a track transit people-mover system and specialist electronic systems.
- Implementation of operational readiness (Oct 07–Mar 08) – ensuring Phase 1 infrastructure was fully complete and that systems were tested, staff trained and procedures in ready for operation in Spring 2008.

Phase 2, which included a second satellite and additional stands, started after 2006 when the residual sewage sludge treatment site was vacated. When completed in 2010, the two phases will enable Heathrow to handle an additional 30 million passengers per year (BAA T5 Fact Sheet: The key stages of Terminal 5).

8.8.1   Project management philosophy

The project management approach on Terminal 5 was developed based on the principles specified in the *Constructing the Team* (Latham, 1994) and *Rethinking Construction* (Egan, 1998) but went further than any other major project. The history of the UK construction industry on large scale projects suggested that had BAA followed a traditional approach, T5 would have ended up opening 2 years late and cost 40% over budget with 6 fatalities; this was not an option for BAA.

Significantly, BAA expected a high degree of design evolution throughout the project in order to embrace new technological solutions and changes in security, space requirements or facilities functionality. On such a complex project, early freezing of the design solution was not realistic.

BAA realised that they had to rethink the client's role and therefore decided to take the total risk of all contracts on the project. BAA introduced a system under which they actively managed the cause (the activities) through the use of integrated teams who displayed the behaviours and values akin to partnering.

This strategy was implemented through the use of the T5 Agreement, under which the client took on legal responsibility for the project's risk. In effect, BAA envisaged that all suppliers working on the project should operate as a virtual company and BAA brought in experienced people from outside the company to spearhead this strategy. Executives were asked to lose their company allegiances and share their information and knowledge with colleagues in other professions. BAA's aim was to create one team, comprising BAA personnel and different partner businesses, working to a common set of objectives.

The T5 Agreement was a unique legal contract in the construction industry – in essence it was a handbook, which provided the appropriate environment for integrated team working. At the high level it is a Delivery Agreement between the suppliers and BAA. This two-page partnering contract, which is executed as a deed is signed by both parties. The following documents are part of the Agreement: the Delivery Team Handbook including the Delivery Agreement; the Project Brief; the Project Execution Plan; the Project Procedures; and the Handbook Data.

The Execution Plan details the project programme, resource and cost plans, team organisational structure, safety processes and outline methodology. Plans are prepared for successive execution phases of the project and require sign off by the suppliers and BAA before further money can be spent. Table 8.1 indicates how the key documents fit together.

The T5 Agreement is based on a cost reimbursable form of contract in which suppliers' profits are ring-fenced and the client retains the risk. It focuses in non-adversarial style on the causes of risk and on risk management through integrated team approaches.

The core values written into the T5 Agreement are *teamwork*, *trust* and *commitment*. A partnering approach was adopted with 80 of BAA's first tier suppliers engaged in the T5 agreement and with BAA acting within the integrated team as project manager rather than simply being the client (BAA, 2007). This approach created an environment in which all team members were equal and problem solving and innovation were encouraged in order to drive out all unnecessary costs, including claims and litigation, and drive up productivity levels (Douglas, 2005).

The reimbursable form of contract meant that there were no claims for additional payments and no payment disputes (NAO, 2005a). BAA used

Table 8.1   Heathrow T5 documents: how they fit together. Source: BAA document

The document:	What it is:
Delivery agreement	The legal deed and conditions of contract.
Supplement agreement	The part, which identifies the skill capability and capacity, the supplier can bring to T5. It defines the potential scope of work on the programme.
Functional execution plan	Details the support required to enable projects to deliver
Sub project execution plan	Details the team's plan of work
Work package execution plan	Details the breakdown of work by each team member/individual supplier (combines preliminaries, specifications & drawings)

Supporting documents:
• Commercial policy
• Programme handbook
• Core processes and
• Procedure
• Industrial relations policy

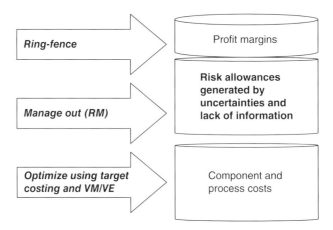

Figure 8.2   Collaborative cost management model. Source: based on Holti *et al.* (2000).

cost information from other projects, validated independently, to set cost targets. If the outturn cost was lower than the target, the savings were shared with the relevant partners. This incentivised the teams to work together and innovate. It was claimed to be the only way to improve profitability; all other costs, including the profit margin, were on a transparent 'open-book' basis (NAO, 2005b). Figure 8.2 illustrates this point. BAA took precautions against risk of the target being too high through a detailed bottom up analysis by independent consultants.

The ring-fencing of profit – the idea that profit is openly declared and paid as an agreed lump sum, disconnected from the cost of labour, materials and plant, is seen by many as one of the most important features of effective SCM. The removal of the incentive for suppliers (in the broadest sense of the term) to *increase* costs to improve profit levels, provided the catalyst for innovation and significant changes in cost levels.

The T5 Agreement created a considerable incentive for performance. If the work was done on time, a third of the predetermined bonus went to the contractor, a third apportioned to BAA and a third went into the project-wide fund that was only paid at Practical Completion of the whole project (Douglas, 2005). Suppliers also benefited from ring-fenced profit and an incentive scheme that rewarded both early problem solving and exceptional performance.

The final strand to the T5 Agreement was the insurance policy. BAA paid a single premium for the multi-billion project for the benefit of all suppliers, providing one insurance plan for the main risks. The project-wide policy covered 'construction all risks' and professional indemnity.

The T5 Agreement allows the project to adopt a more radical approach to the management of risk including early risk mitigation. Key messages include: *'working on T5 means everyone anticipating, managing and reducing the risks associated with what we're doing'* (T5 Agreement,).

8.8.2   Best practice on T5 – The Mechanical and Electrical (M&E) Buy Club

The concept of a Buy Club, which was pioneered on the T5 M&E trades, created a new era of openness and collaboration with the potential for the elusive *world-class* results. The M&E Buy Club pooled the expertise and buying power of three 1st tier M&E contractors and sourced each of thirteen specialisations from (generally) one supplier who was then responsible for supplying all sixteen projects at T5. The outcome of this innovative strategy ensured a consistent approach on the whole terminal to the £600m spend on the M&E equipment and materials. It ensured that the specifications addressed the quality and the life cycle actually required and it ensured a best value solution.

Other critical benefits included:

- Cost: a 10–30% cost saving to BAA on the budget for M&E materials and equipment;
- Time: the Club appointed the suppliers early, engaged them in design and promoted lean manufacturing and installation;
- Quality: early agreement on benchmark prototypes with an open book approach which revealed issues before they became problems;
- Safety: planning by the Buy Club reduced unsafe working;
- Offsite manufacture of modules reduced assembly hours on site;
- Logistics: simplifying the supply chain made logistics much easier to manage;
- The environment: the tender assessment score included 5% for environmental issues (e.g. recycling of materials).

The M&E Buy Club was deemed so successful for the M&E package of work that BAA used the same process for the £200m fit-out and the £50m communication systems packages.

Part B

The main differences between the BAA T5 procurement approach and more traditional approaches was that the Buy Club did not insist on a single source supplier and sensible alternatives were not excluded. Packages worth more than £250k had a unique acquisition plan (the same as all second-tier procurement on the project); bids were submitted on a lump sum basis for benchmark designs, but open-book costing applied when design development occurred and all bidders were invited to the common briefing.

8.8.3   Best practice on T5 – construction logistics consolidation centres

BAA first pioneered the use consolidation of centres with the Logistics Consolidation Centre at Hatton Cross, Heathrow airport in November 2001. At the time, this was a revolutionary concept and unproven in the construction industry. The Centre attracted a great deal of interest and became an M4i (www.m4i.org.uk) demonstration project and the test bed for the huge T5 consolidation centre at Colnbrook.

At Hatton Cross, the aim was to store goods for no more than seven days with the goods delivered to the workface when required. This approach reduced vehicle movements on site, minimised the storage and reduced damage and waste with gains up to 5% recorded (www. constructionexcellence.org.uk – cited 15 November 2007).

The T5 site, despite being one of Europe's largest infrastructure projects, was physically constrained and as a result space for construction activities was at a premium. Added to this there was a need to minimise construction traffic on local roads.

In response to these requirements BAA created two consolidation centres. The Colnbrook Logistics Centre, which provided three principal areas: a railhead that brought in the bulk materials from Europe and the UK; a factory for the prefabrication and assembly of steel reinforcement, and a 'lay down' area. The second distribution centre, the Heathrow South Logistics Centre, initially provided a facility for the automated manufacture of pile reinforcement cages and later acted as a centre where materials were assembled into work packages ready for delivery to site (BAA Heathrow T5 Fact sheet).

To effectively control the delivery process, a logistics strategy was developed by BAA, which resulted in several benefits including:

- eliminating the need for lay down space for materials;
- increasing the reliability and efficiency of supplied materials, which in turn increased the productivity levels from an average of 55–60% to an unprecedented 80–85%;
- reducing the waste created by traditional practices;
- reducing transport movements enabling BAA to uphold environmental commitments.

Project teams were required to plan their requirements for materials up to six weeks in advance. Supporting this process was a software system specifically designed for the construction industry called *Project Flow*, which collated the team's demands and drove the materials through the system to enable delivery on site just before or on the date required. This *just-in-time* or *pull* strategy, which was based on the methods used in factory based manufacturing, was a first for a construction programme of this scale. It was supported by the extensive use of prefabrication and pre-assembled components (BAA Heathrow T5 fact sheet).

8.8.4   Best practice on T5 – 3D modelling with a single model environment

BAA set itself a target of using technology to reduce the total project cost of Heathrow Terminal 5 by 10 per cent. This was largely achieved by creating a single 3D computer model that BAA and its project partners used to design, build and ultimately maintain the terminal building.

Major projects, involving multiple design teams, frequently suffer from poor collaboration and ambiguous design detail resulting in delays and increased costs. By providing a single model environment (SME) this provided a single solution so that all framework partners could collaborate. Information contained within the model was used to plan the methodology of constructing the building, to manage time and to improve health and safety – even to plan the retailers' fit-outs. Anyone could use the model without specialist CAD knowledge, as it could be viewed and information extracted using the common 3D viewer – *NavisWorks*.

BAA used Autodesk® *Architectural Desktop*™ as its CAD platform and *Documentum*™ as its document management platform. 3D graphics files were converted by the program *NavisWorks* into a format (.nwd.format) that enabled clash detection software to be run and renderings, images, fly-throughs and animations, to be produced, as well as 4D construction planning. 4D construction planning is a work planning process that adds time as a fourth dimension to programmes with CAD data (2D or 3D), creating real-time graphical simulation of planned works (Beardwell *et al.*, 2006).

Conventionally, the architect designs the building and passes the CAD drawings over to the engineer. The engineer then draws the building over again for engineering analysis, as do the subcontractors and the result is that the building and the elements within it are redrawn hundreds of times. BAA has proved in its research into the construction process that by the time the project gets to site these drawings are bound to contain inconsistencies, meaning that if different parts do not fit together they have to be reworked on site. The estimated cost of wasted time and materials alone is at least 10 per cent of total project cost. If the costs of disruption to the programme are factored in, the figure is even higher.

The idea behind the single project model is to derive an unambiguous set of data through the sharing of data. Using this approach engineers never redraw the information; they re-use the architect's data and add to it. This approach drives out errors and improves efficiency.

A 3D model incorporating intelligent object technology was used at T5 to improve efficiency even more. This meant that objects in the CAD drawing *knew* what they were, and how they fitted with other objects in the building.

The massive roof nodes connecting the roof structure were a good example of how the single-model environment worked in practice. Richard Rogers Partnership (RRP) designed the node and passed it over to structural engineer Arup. The engineer used the architect's drawing to carry out structural analysis; RRP then modified the design to fit the analytical requirements using the same set of data. The model was passed to the steel fabricator Rowen Structures who used it to fine-tune the design of the parts of the node that had to be specially made. Finally, they used the model to control the machinery that made the roof parts (Pearson, 2003).

The big challenge on the T5 project was to ensure that everyone was collaborating, not just communicating but sharing information. Mervyn Richards, CAD Technology Manager at Laing O'Rourke, one of BAA's framework advisers considered that 'The use of technology on this project has enabled a full collaborative environment with all of BAA and its 42 Framework partners' (http://www.navisworks.com/en/solutions/casestudies/baa – cited 5 November 2007).

Lessons learned at T5 have been disseminated to the rest of industry – The Construction Project Information Committee has published the *Code of Practice for Production Information*, which contains the processes and protocols used at T5 (CPIC, 2003).

8.8.5   Best practice on T5 – value engineering

As T5's main roof was a large element in the structure, designing a cost effective solution was critical to the project's success. Richard Roger's Partnership's competition winning design envisaged a glorious waveform roof supported on four rows of branched structural columns. This proved to be too complex and beyond the capability of the contractors. It was also deemed too expensive for the client, BAA.

In December 1999, a major value engineering exercise was undertaken involving all the key players: architects RRP, structural engineer Arup, steelwork contractor Severfield-Rowen, cladding specialist Schmidlin and Hathaway Roofing.

The development of the successful design became something of a saga with a solution developed through an iterative process. Buildability was a major issue due to the restrictions on site – at its highest point the roof towers 37 m above the apron, however the airport's radar is in operation 2 m above that thus prohibiting the use of cranes.

In the end, the design team came up with a solution that satisfied all criteria: a single span tied (or bowstring) arch supported high above the concourse on inclined structural columns. The roof was assembled on the ground in bays using 3000 preassembled cassettes. The bays were then jacked up using the support abutments – in all, five lifts of three bays each and one single-bay lift (Pearson, Building 2003).

To minimise any chance of mishaps and to ensure that the roof erection proceeded smoothly on site, the T5 roof team (including designers, suppliers and fabricators) pre-erected one of the twenty two major roof abutment structures at the steel fabricator's (Severfield-Rowen) base near Thirsk in Yorkshire.

The pilot exercise proved to the design team that the erection method was workable and helped the construction team better understand the sequencing and tolerances required. As a result, the T5 team identified 140 significant lessons resulting in each having a risk mitigation plan enabling faster construction on site.

This exercise cost BAA £4 m, but saved three months work on the Heathrow site, enabling delays that had previously arisen due to the wet winter of 2001/02 to be recovered (NAO, 2005b). This is another classic example of best practice on the T5 project – proactive risk management using an integrated team approach.

8.8.6   Best practice on T5 – offsite pre-fabrication

BAA has long considered that off-site manufacture is the way forward on their construction projects. In 2002 Tony Douglas, BAA's Group Supply Director, introduced a very ambitious target of 65% of every project being preassembled (BAA, 2002).

With so little space on site, off-site prefabrication was an important element in the success of the T5 project. Off-site manufacture has many advantages; for example the components can be assembled in clean, efficient, safe and secure factories, which improves the quality of assembly and reduces the amount of materials wasted and stolen. A further benefit is that prefabrication reduces the pressure on the labour market in the south-east of England. This, in turn, should lead to faster build times and increased productivity. In fact, BAA estimated prefabrication has led to an increase in productivity of between 10% and 15% compared to the average building site (Building Magazine Supplement, 2004, p40).

An example of off-site fabrication on T5 was on the M&E services where Amec Services manufactured and assembled 60% of the services offsite based on a modular services system. More than 5,000 modules based on 11 standard types were supplied to the main terminal building. Overall, AMEC used less than half the site labour and yet, at the same time, reduced the project's build programme by six months.

8.8.7   Best practice on T5 – project control system

The T5 project aimed to be in the forefront of project control and was one of the first major users of the *Artemis* project management system in UK construction. The system is a very robust and can show how each area of the project is performing relative to target, on both schedule and costs.

Part B

A further key point of the *Artemis* system is that it can give information at programme or at individual project level or sub-project level. Cost and performance data can be analysed in various ways including the production of two highly useful indices, the *Schedule Performance Index* and the *Cost Performance Index*, which are generated for all the levels and for each package (Hill, 2004).

Under a £250m contract Vanderlande Industries were required to provide a state-of-the-art baggage handling system at Heathrow T5. Wickramatillake *et al.* (2007) identified some of the difficulties of accurately calculating earned value analysis performance measurement on this complex major sub-project. Eight areas of concern, with recommended solutions for improving performance measurement calculations, were outlined. However, overall, the performance measurement methodology was considered a success, and Vanderlande stated that they intended to use this methodology for all medium and large projects in the future.

8.9   Conclusions

The journey from the potentially disastrous Heathrow Express project in 1993 to the successfully completed Heathrow T5 in 2008 has been a massive challenge for all. BAA identified that successful delivery was all about leadership and culture and as an expert client, over the last decade or so, has developed a radical new approach to project management.

We have seen how the process has developed initially through partnering agreements and, latterly, through the innovative T5 Agreement. Throughout this last decade BAA have been at the forefront in developing the use of modern procurement methods through the appointment of integrated supply chains in which the parties have the long-term objective to work together to deliver added value to the client.

Under the T5 Agreement, BAA accepted all the risk from the outset and guaranteed its suppliers an agreed margin. The Agreement asked their supply chain partners to demonstrate commitment, trust and team work over a long term, in return for a guaranteed margin with the potential for earning bonuses if performance improvement could be demonstrated.

The T5 approach created an environment in which all team members were equal and were required to work in integrated teams. Furthermore, it encouraged problem solving and innovation in order to drive out all unnecessary costs, including claims and litigation, and drive up productivity levels. The big challenge on the project was *to harness the intellectual horsepower* in order to create the *World Class Performance* demanded by Sir John Egan.

BAA's enlightened approach created a collaborative environment, which led to the implementation of industry best practices and world-class performance. This approach is particularly relevant to long-term projects with high risk and high complexity, valued at £200 million and above, but might not be so relevant for smaller, more straightforward projects.

So did Terminal 5 represent history in the making? In his 2005 lecture at the Royal Academy of Engineering, Andrew Wolstenholme, BAA's Project Director, confirmed that he believes it did and that the new approach to project management as set out in the T5 Agreement will help the industry change for the better. However it will require a massive culture change to become the norm.

8.10   Acknowledgements

The author would particularly like to thank the following people who have contributed to the production of this chapter:

- Dr Stephen Pryke at UCL for generously loaning some of his PhD source material which included information on the Genesis Project.
- Antonia Kimberley, Head of Media Strategy, BAA Heathrow and her colleagues at BAA who checked the final draft for errors and omissions.

References

BAA (1997) *In Context*, 'Genesis', Issue No. 5 December 34–35.

BAA (2002) *In Context*, 'Time is Money'. Issue No. 21 Interview with Tony Douglas, 8–9.

BAA (2007) *An official T5 report for the aviation community*, Part two, PPS Publications.

Beardwell, G., Honnywill, T., Kainth, T. and Roberts, S. (2006) Terminal 5, London Heathrow: 3-D and 4-D design in a single model environment. *The Arup Journal* 1, 3–8.

Bennett, J. and Baird, A. (2001) *The NEC and Partnering: the guide to building winning teams*. Thomas Telford, London.

Brady, T., Davies, A., Gann, D. and Rush, H. (2006) Learning to manage mega projects: the case of BAA Heathrow Terminal 5. *IRNOP VII Project Research Conference, October 11–13*, Xian, China, 455–467.

Building Magazine (2004) Terminal 5 Supplement. A Template for the Future. How Heathrow Terminal 5 has rebuilt the building industry. In *Building*, 27 May.

CPIC (2003) *Code of Practice for Production Information*, Construction Project Information Committee, Alton, Hants.

Constructing Excellence (2004) T5 Buy Club: How M&E contractors pool purchasing at Heathrow Terminal 5, http://www.constructingexcellence.org.uk/pdf/case_studies/t5_buy_club_20040916.pdf – accessed 6 September 2007.

Cox, A. and Townsend, M. (1998) *Strategic Procurement in Construction: Towards better practice in the management of construction supply chains*, Thomas Telford, London.

Douglas, T. (2005) Interview: Terminal 5 approaches take-off. In *The Times*, Public Agenda Supplement, 6 September.

Egan, J. (1998) *Rethinking Construction: The Report of the Construction Task Force to the Deputy Prime Minister, John Prescott, on the Scope for Improving the*

Quality and Efficiency of UK Construction, Department of the Environment Transport and Regions Construction Task Force, London.

Finch, A.P. and Patterson, R.L. (2003) *Recent trends in procurement of tunnel projects in the United Kingdom*, www.tunnels.mottmac.com/files/page/1607/ Procurement_of_Tunnel_Projects_in_UK.pdf.

Rapid Excavation and Tunnelling Conference (RETC) Proceedings (2003) (www. tunnels.mottmac.com/files/page/1607/Procurement_of_Tunnel_Projects_in_UK. pdf – cited 20 September 2007).

Gray, C. (1996) *Value for Money*, Reading Construction Forum and the Reading Production Engineering Group, Berkshire.

Hill, A. (2004) Follow the Money in *New Civil Engineer*, June p. 20.

Holti, R. (1996) *Designing Supply Chain Involvement for UK Building*, Programme for Organisational Change and Technological Innovation, The Tavistock Institute, London.

Holti, R., Nicolini, D. and Smalley, M. (1998) *Prime Contractor's Handbook of Supply Chain Management*. Tavistock Institute, London.

Holti, R., Nicolini, D. and Smalley, M. (2000) *The Handbook of Supply Chain Management*. Construction Industry Research and Information Association, London.

Latham, Sir M. (1994) *Constructing the Team: Joint Review of Procurement and Contractual Arrangements in the United Kingdom Construction Industry*. HMSO, London.

Lownds, S. (1998) 'Management of Change: Building the Heathrow Express Leveraging team skills to get a business rolling – the story of the change of culture on the construction of the Heathrow Express Railway.' Paper presented to Transport Economists' Group, University of Westminster, 25 November 1998.

Mace (1997) *Genesis HAL World Cargocentre MSCP World Class Project Process: The Result*, Mace Limited, London.

MecLeod (1999) "Supplier Development Workshop", BAA in house magazine. In Contest, Issue 10, April p. 7.

MPA seminar (2000) "Project Recently: Breaking the Cycle of Failure", held at Royal College of Pathologists, London, June. http://www.majorprojects.or/ public1631.pdf.

National Audit Office (2005a) *Improving Public Services through better construction*, HMSO, London.

National Audit Office (2005b) *Improving Public Services through betterconstruction*: Case Studies, Report by the Comptroller and Auditor General | HC 364-II Session 2004–2005 | 15 March, HMSO, London.

New Civil Engineer (1997) Common Interest within 21st Century Airports. Published as a supplement to *New Civil Engineer* in association with BAA, pp 9–11, Emap Construct.

New Civil Engineer (2004) Terminal 5 Supplement, published as a supplement to *New Civil Engineer* in association with BAA, February, Emap Construction network.

Partnership Innovation Initiative (2004) Construction Logistics Consolidation Centres: An examination of New Supply Chain Techniques – Managing and Handling Construction Materials.

(www.constructingexcellence.org.uk).

Pearson, A. (2003) T5 Satisfying hells hounds wrestling with serpents. *Building* 25 July.

Pryke, S.D. (2001) *UK Construction in Transition: Developing a Social Network Analysis Approach to the Evaluation of New Procurement and Management Strategie.*, PhD thesis, The Bartlett School of Graduate Studies, University College London.

Sauer, G. (1994) "Collapse Chaosat Heathrow", Contract Journal, 27 October.

Wickramatillake, C.D., Koh, S.C.L., Gunasekaran, A. and Arunachalam, S. (2007) Measuring performance within the supply chain of a large scale project. *Supply Chain Management: An International Journal*, 12(1), 52–59.

Womack, J.P. and Jones, D.T. (2003) *Lean Thinking: banish waste and create wealth in your corporation* 2nd edition, London: Simon and Schuster.

Womack, J.P., Jones, D.T. and Roos, D. (2007) *The Machine that Changed the World: How Lean Production Revolutionized the Global Car Wars*. London: Simon and Schuster.

Part B

9

Supply Chain Management: A Main Contractor's Perspective

Andrew P. King and Martin C. Pitt

Andrew P. King and Martin C. Pitt

9.1 Introduction

The need to utilise Supply Chain Management (SCM) as a way of improving the project delivery process has become one of the dominant best practice messages in the UK construction industry over recent years. However, there is a clear gap between best practice advice and the reality of the business world. The advice available is far too generic to be practically employed by practitioners seeking to implement SCM. Added to this problem is the overriding myopic tendency towards considering SCM from the client's perspective. Despite the fact that main contracting organisations have such an important role to play in channelling client demand through their own supply chains, these organisations are overlooked when it comes to practical useful advice supported by rigorous empirical research. This chapter is aimed at fulfilling this need by exploring SCM from a main contractor's perspective using action research based on a case study of Morgan Ashurst (formerly Bluestone) plc, a major nationwide UK main contractor. The case study provides an overview of the SCM development process from a variety of different perspectives, including the contractor's internal staff and its subcontractors. The results are presented thematically and, as they offer a degree of generalisability, are particularly useful to other main contractors faced with developing their own approaches to SCM.

9.2 Supply Chain Management

'Supply Chain Management is the integration of key business processes from end user through original suppliers that provides products, services, and information that add value for customers and other stakeholders' (Lambert *et al.*, 1998).

This definition of SCM, developed in conjunction with the Global Supply Chain Forum, highlights the importance of integrated business processes and customer focus. Customer focus is important because it draws attention to the way the chain is market-driven rather than supplier-driven (Christopher, 1998). The concept of SCM has grown in popularity in recent years as organisations seek to increase competitive advantage. Increasing adoption of SCM has been accompanied by a growing shift in focus from price reduction to value creation. For example, Drayer (1999) has reported how Proctor and Gamble has begun to focus on utilising SCM to create value for the entire supply chain, as the company could no longer guarantee success from its traditional focus on quality products and innovative customer-focused marketing.

9.2.1   Supply chain management in construction

Although lagging behind many industries, construction has seen increasing calls for the implementation of SCM to increase performance through integrated project delivery processes (Latham, 1994; DETR, 1998; Strategic Forum, 2002; CBPP, 2003). However, the need to question the increasing importance of SCM has been raised by authors such as Green (1999) who believe that much of the agenda is awash with rhetoric and dogma. Green states that:

> . . . less scholarly 'best practice' literature frequently ignores the structural barriers to SCM, preferring to concentrate on the need for 'culture change' (2005: 579).

Deepening the critical debate in this way serves a useful purpose as it helps avoid blind adoption of the latest faddish management thinking. The majority of supply chain literature is characterised by such vague concepts as culture, collaboration, integration and relationships. Mouritsen *et al.* (2003) have cautioned against the universal promotion of such concepts as integration and collaboration, without taking account of the supply chain environment and specific power relationships. They argue that 'best practice' in SCM 'should only be copied if the objective situational factors are exactly the same, which is very seldom the case' (Mouritsen, *et al.*, 2003: 694).

The importance of relationships in buyer-supplier exchange has a long history (see for example, Poirier and Houser, 1993; McHugh *et al.*, 2003; Bullington and Bullington, 2005). Blake and Mouton's Dual Concern Model (1964), shown below in Fig 9.1, is particularly interesting as it simply describes common types of exchange behaviour in construction. By focusing on the concern two parties have for each other, it shows how having a high concern for one's own interest and a high concern for the other exchange party's interest correlates with compromise (Box D), where both parties win a bit and lose a bit; hardly an ideal state. On the other hand this might be regarded as appropriate in a project environment.

Cox (2004a) and Cox, *et al.* (2007) take the view that win–win outcomes are not feasible in relationships between buyers and suppliers in exchange transactions as all exchanges are contested. Cox *et al.* (2007: 278) make

		Concern for other's interest	
		Low	High
Concern for own interest	Low	**A** Avoidance *Both parties lose*	**B** Forbearance *One party loses and one party wins*
	High	**C** Rivalry *One party wins and one party loses*	**D** Compromise *Both parties win 'a bit' and lose 'a bit'*

Figure 9.1   Dual concern model. Source: adapted from Blake and Mouton (1964).

their views clear that 'any attempt to search for win–win outcomes is a waste of everyone's time and effort, whether in construction or in any other types of supply chain or market'. Cox (2004a, b), Cox and Ireland, (2002) and Cox, *et al.* (2000, 2004, 2007) consider the UK construction industry from a power and leverage perspective of relationship and performance management. Their development of a theoretical framework, which takes account of the structure of the industry, buyer and supplier power and leverage and its effect on appropriate relationships, has provided much needed clarity for those struggling to get to grips with the implementation of SCM systems. One of their main arguments, that few clients have sufficient standardised long-term demand to develop a highly collaborative partnered supply chain, has significant implications for the client-focused nature of SCM in the UK construction industry.

9.2.2   Client or contractor-led supply chain management?

The body of Cox-led work referred to above makes interesting reading, but a very much client-centric focus has dominated the discussion of SCM in the UK construction industry. Many influential reports have stressed that clients need to be at the centre point of the supply chain (Latham, 1994; DETR, 1998; Strategic Forum, 2002; Briscoe *et al.* 2004). Briscoe *et al.* (2004), picking up on this common theme, found that clients are the most significant factor in achieving integration of the supply chain. More specifically, they found that without the client's desire to develop supply chain relationships, integration could not be achieved. Taking account of the aforementioned views on the limited amount of clients who can create the appropriate power and leverage environment, where does this leave the rest of the industry?

9.2.3   Organisational and project supply chains

Male and Mitrovic (2005) help answer this question by offering a useful way of thinking about supply chains by distinguishing between the *types* of supply chain. Building on Male's (2002) earlier airport and airline analogy, which saw the main contractor as the supply chain network 'hub' meeting various client needs by managing various project-specific supply chains, they draw a distinction between the project supply chain (PSC) and the organisational supply chain (OSC). The PSC directly relates to a specific client requirement, whilst the OSC relates to the main contractor's organisational supply chain (Male and Mitrovic, 2005). The notion of an OSC is particularly interesting as it draws attention to the main contractor's ability to manage and influence a number of project-specific supply chains for a number of different clients, irrespective of the client's inclination and ability to utilise SCM. In contrast to the majority of client-centric SCM literature, which struggles to find ways to place clients in the role of chief protagonists, this distinction recognises that main contractors with sufficient organisational and economic size, as the hub of numerous supply networks, have the ability to develop their organisational supply chain and provide numerous highly differentiated clients with the resulting benefits. With the exception of large construction clients like British Airports Authority (see also Chapter Eight) and the Ministry of Defence, few clients have sufficient repeat demand to develop their own SCM approaches. Sustained high levels of demand are needed to maintain *standing supply chains*, whose members are willing to invest and innovate for the benefit of a single client.

For Morgan Ashurst, the OSC is especially pertinent. The company's position as a large-scale nationwide contractor, working on an average of 150 projects at any one time for a number of different clients with diverse needs using a range of different procurement routes, means it recognises the difficulties that its clients experience in struggling to make SCM provide them with tangible benefits. The main contractor's ability to form long-term relationships with subcontractors stems from its ability to provide a multitude of clients with the benefits of an OSC, irrespective of the client's disposition towards SCM. However, taking account of the client-centric focus of SCM, it can be argued that the main contractor's ability to act as a 'demand channel' between clients and subcontractors is not made sufficiently clear in the SCM literature. This focus may lead to a failure in making clear the danger of opportunism in main contractor-subcontractor relationships:

> *Opportunism is a rational response for those involved in one-off games, in which there are no incentives for higher rewards from not maximising returns in the short-term. Obviously, collaboration is a better alternative if there are incentives that allow parties to the exchange to envisage higher returns rewards in the future. In such circumstances maximising short-term advantage is not a logical response to the superior commercial opportunities that may be feasible in the future from entering into bilateral dependency operationally* (Cox et al., 2007: 31).

Whilst recognising that collaboration is a preferable option in certain circumstances, it is not made sufficiently clear that main contractors and their subcontractors are in a prime position to benefit from such proactive, collaborative, incentivised, long-term partnered approaches, providing that

Part B

such firms have the appropriate attributes and approach. By focusing on the inappropriateness of a long-term partnered approach for the majority of UK construction clients, owing to their lack of ability to provide sustained standardised long-term demand, Cox's *et al.'s* work (2007) is similar to the majority of UK SCM literature in failing to stress how main contractors, with the necessary downstream demand, can propagate SCM and share the benefits with their clients and supply chain. It is acknowledged that Prime Contracting through the 'Building Down Barriers' initiative of the Ministry of Defence (Holti *et al.*, 2000), took a contractor – centric position placing, intentionally, the contractor in the position of supply chain manager. This public sector initiative was ambitious but not universally adopted even with the Ministry of Defence itself.

9.2.4   Contractor and subcontractor relationships

Despite the opportunities available to some main contractors, the difficulty involved in developing a managed supply chain is all too apparent. Ireland (2004) outlines some of the main problems. As the integrator of numerous supply chains, contractors need to get a regular profitable workload whilst managing a supply chain of subcontractors and suppliers who are fighting for their own survival in an environment where adversarial relationships and opportunism are the norm, as low barriers to entry maintain the fragmentation and low levels of profitability. Previous work focusing on supply chain relationships and 'partnering' paints a similarly difficult picture. For example, Dainty *et al.* (2001) explored subcontractors' perspectives of supply chain alliances and found that 'there remains a general mistrust within the SME (small to medium-size enterprises) companies that make up the construction supply chain, and a general lack of belief that there are mutual benefits in supply chain integration practices' (Dainty *et al.*, 2001: 847). Briscoe *et al.* (2001) found attitudinal barriers to collaboration at the subcontractor-main contractor interface, whilst Muya *et al.* (1999) drew attention to poor practices such as late payments and a lack of feedback. Briscoe and Dainty (2005), in their case studies of three construction clients, found integration of the supply chain was inhibited by a lack of sufficient trust to allow formal partnering to endure.

9.2.5   Advice for those seeking to implement SCM

Some of the problems identified above are the legacy of the dominance of adversarial relationships between main contractors and their subcontractors in the UK construction industry. However, they are also perhaps a result of the generic advice available to practitioners, which often lacks detailed information explaining how to implement construction-specific SCM initiatives (Christopher, 1998). This is a particular issue because SCM has its origins in non-construction sectors; attempts to adopt models from other industries, without recognition of context, are fraught with difficulties

(Fisher and Morledge, 2002). Green *et al.* (2005), recognising that there has been a blind justification for construction to emulate SCM from other industries without comparing industrial contexts, compared the aerospace and construction industries and stressed the need to recognise the differing inter-sector focus of SCM in addition to organisational context. It is worth highlighting that, even without such a debate, practitioners wishing to develop their own SCM approach still encounter problems as they search for practical advice demonstrating how to turn the sentiments of generalised, non-sector specific, best practice into a reality.

9.2.6   Summary on the principles of supply chain management

Effective SCM requires a carefully considered approach that is clearly aligned with industry, client, supplier and organisational specifics. This aspect of 'fit for purpose' was deemed to be one of the most important aspects of Morgan Ashurst's strategy, instead of trying to utilise a predetermined supply chain model the team sought to build their approach from first principles (Cox and Thompson, 1998; Cox and Townsend, 1998).

In drawing attention to the key role that main contractors have to play in SCM, the preceding journey through construction-related SCM literature has clearly highlighted that while a client-centric focus still pervades, some main contractors are in a position to make SCM a reality. Studies exploring the main contractor-subcontractor supply chain relationships highlight the problems with current practice; problems perhaps linked to the lack of context-specific practical advice grounded in empirical evidence available for those wishing to develop their own approach to SCM. This chapter seeks to address the need for practical advice through the use of an action research case study of a main contractor developing its approach to SCM.

9.3   Methodology

This action research follows the development of one organisation's subcontractor supply chain and concentrates on the early development of an overall strategy which took place during 2006. Action research (involving action-oriented and participatory approaches), develops 'practical solutions to issues of pressing concern to people' by joining action, reflection, theory and practice (Reason and Bradbury, 2001: 1). This is particularly effective in an environment of organisational change where it can help develop effective work practice (Coghlan and Brannick, 2004). The research uses a modified grounded theory methodology (Glaser and Strauss, 1967) which recognised prior knowledge of significant issues relating to Morgan Ashurst's subcontractor supply chain (Strauss and Corbin, 1998). These issues were used as a starting point to structure the research, and their importance to the study was tested through the accumulation of new data. This approach stands in contrast to the more inductive approach outlined by Glaser in his later work, *Basics of Grounded Theory Analysis: Emergence vs. Forcing* (1992).

The *a priori* and emergent issues were contextualised through a detailed ongoing literature review which explored SCM, both in construction and other industries.

The work is based around a case study of a major nationwide construction contractor, Morgan Ashurst. The case study included semi-structured interviews with 48 participants; the majority were Morgan Ashurst representing various geographic and functional areas in the business, but representatives from subcontracting organisations with whom they worked were also included. The majority of the interviews were fully transcribed and analysed using Version 7 of QSR's Nvivo qualitative data analysis software.

In addition to the semi-structured interviews, the case study included document analysis and participant observation of a series of supply chain seminars undertaken by Morgan Ashurst throughout the country during 2006. The seminars were attended by subcontractors with whom Morgan Ashurst had previously worked and developed relationships. The company provided its subcontractors with the opportunity to comment on its initial plans for developing the supply chain, as well as offering them the opportunity to express their views about working with Morgan Ashurst. In addition to treating the seminars as a data-gathering exercise, they were used as a way to increase 'buy-in' and adoption of the new approach to Morgan Ashurst's subcontractor supply chain. Morgan Ashurst also formed a national working party whose task it was to develop the company's supply chain vision into specific tools and techniques to underpin the approach. Once more, participant observation and document analysis were used to incorporate the working party into the study.

9.3.3   Organisational setting

The process that Morgan Ashurst followed to develop its subcontractor supply chain is shown below:

- Develop a process to understand the current state of the company's subcontractor supply chain including:
  - Consultation – both internally with Morgan Ashurst staff and externally with its subcontractors;
  - Internal studies comprising: 1) analysis of subcontract orders; 2) internal surveys; 3) internal key performance indicators (KPIs).
- Consider supply chain management and relationships in detail in order to inform the development of the supply chain:
  - Conduct a detailed literature review;
  - Attend supply chain seminars.
- Develop an overarching strategic vision for the future state founded on empirical evidence and structured around the principles of:
  - Relationships;
  - Culture;
  - Consolidation;

- ○ Consistency;
- ○ Cost;
- Disseminate the findings:
  - ○ Publish an internal paper entitled 'Defining the Vision', outlining the overall strategic vision.
- Carry out further consultation by:
  - ○ Holding interactive Morgan Ashurst and subcontractor seminars;
  - ○ Forming a nationwide Supply Chain Working Party, and a number of similar business unit based groups, consisting of various members of Morgan Ashurst staff from a variety of geographical and functional parts of the business.
- Disseminate the detailed findings by publishing a second paper entitled 'Realising the Vision' which sets out the following detailed changes:
  - ○ The 5-step approach to developing business unit based supply chains;
  - ○ The Supply Chain Principles: a hierarchy of subcontractors with associated features and benefits, Principles of Engagement and Top 10 Behavioural Tips.

9.4   Analysis

This section outlines the five dominant themes that emerged during the study (Figure 9.2):

- Relationships
- Culture
- Consolidation
- Consistency
- Cost

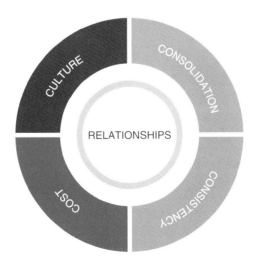

Figure 9.2   Morgan Ashurst's 4Cs and R Model.

Part B

9.4.1   Relationships

The importance of relationships grew throughout the study until it developed into the 'core category'. The approach of engaging with internal staff, and externally with subcontractors, was deemed to be one of the most important aspects of the study. Interestingly, both staff and subcontractors involved in the case study stated that whilst they had been involved in other main contractors' development of SCM, this was the first time they had been actively consulted for their views. Further, they stated how favourably they viewed such engagement and responded by declaring a strong commitment to working closely with Morgan Ashurst to develop the supply chain.

Whilst Morgan Ashurst's business benefited from many strong and successful relationships developed over time with subcontractors, in some areas these relationships remained short term and were originated through 'lowest tender price' bidding. There was an overriding belief that the business would be better served by cultivating and protecting healthy relationships in order to provide mutual benefits. Similarly, although the performance of Morgan Ashurst's subcontractors was informally recognised and rewarded, there was a failure to formally recognise these relationships in order to generate additional benefits for both parties. Notwithstanding the variety of strengths characterising the company's relationships with its subcontractors, no particular incentives or differential in status or trading terms were being offered.

Other areas deemed significant included a lack of subcontractors' early input into tenders. Despite parts of the business benefiting from the increased competitiveness and risk reduction that gaining subcontractors' early involvement in scheme development provides, there was a failure to develop relationships and systems that consistently guaranteed such input. Similarly, despite experiencing the power of introducing clients, at tender stage, to the subcontractors that would be working with them on their project if Morgan Ashurst were selected, the ability to demonstrate a managed supply chain to the external market was not uniformly practised across the business.

Long-term relationships and mutual wins

Participants cited many examples where relationships defined by collaboration, integrity and a concern for each other's interests led to both parties in the exchange gaining substantial increases in performance such as increased quality, lower programme periods, reduced cost growth and indeed reductions in cost in addition to putting the project at the centre of the table as opposed to different organisations. The findings are more closely associated with a relational perspective to buyer and supplier exchange, and as such can be contrasted with the power and leverage perspective adopted by Cox (2004a, b), Cox and Ireland, (2002) and Cox *et al.* (2000, 2004, 2007). They argue that although 'win–win' is an appealing idea, it has no basis in reality when one adopts an economically rational view of commercial exchange between buyers and suppliers. They state that this is because positive sum win–win outcomes are not objectively feasible; only non-zero sum mutually beneficial restricted forms of mutuality are available. Further, they state that a desire to achieve outcomes that are beneficial to both parties could show a serious misunderstanding of buyers and supplier exchange. In simple terms, their argument is founded on an economically rational view of economics. However, approaching win–win from a different perspective – that of a socially-constructed reality (as with this research),has shown that participants firmly believe that where both parties show a concern for each other's interests, they both derive financial and non-financial benefits.

Becoming 'bogged down' in seeking an objective 'win', or indeed becoming overly concerned with the equal sharing of that win is of secondary importance to the practitioners involved in this study; they know from experience the general increase in benefit that can be realised by numerous parties sharing a common goal. Objective arguments against the vagueness of win–win outcomes offer little if one accepts that the business world is socially constructed and that the term 'client' generally represents numerous different stakeholders – each with potentially competing and dynamic needs. It follows that achieving an objective equally shared perfect win is highly unlikely as it is neither clear what exactly is being optimised, nor likely that an optimal outcome exists. Nevertheless, this does not stop 'win–win' being a highly useful concept for the pragmatic organisation that recognises it is dissatisfied with current performance and wishes to use this concept to act as a powerful motivational instrument. Responding to these observations led Morgan Ashurst to modify Blake and Mouton's model (1964) and incorporate this as an essential part of its supply chain vision, as shown in Fig 9.3. In taking such an approach, Morgan Ashurst has excluded a strictly

Triple Concern Model		Concern for other's interest		
		Low	High	Very High
Concern for own interest	Low	**A** Evasion (We both lose)	**B** Powerlessness (I lose, you win)	
	High	**C** Rivalry (I win, you lose)	**D** Compromise (We both win a bit and lose a bit)	
	Very High			**E** Win-Win-Win (We all Win) FOCUS ON THE CLIENT'S CONCERNS

Figure 9.3 Triple-concern model. Source: adapted from Blake and Mouton (1964).

rational economic view and associated objective win that can be equally shared; rather the company draws on its own experiences which show that where having a very high concern for one's own interest, one's subcontractor and one's client can lead to a 'win' for each party. This approach purposefully removes the negative terminology of both winning and both losing 'a bit'. In doing so, it takes account of the need to develop long-term relationships underpinned by the following properties:

- Trust
- Communication
- Interdependence
- Collaboration
- Commitment
- Integrity and honesty
- Flexibility
- Concern for each other's interests.

9.4.2 Culture

Despite developing informal long-term relationships, the lack of a structured managed supply chain consistently utilising the same subcontractors was found to lead to an inability to develop a shared culture. As such, awareness of and commitment to key business drivers contained within Morgan Ashurst's business improvement strategy was not consistent. There was widespread agreement of the need to change the way Morgan Ashurst interacted with its subcontractors and promoted core business objectives. As such, the resulting MASC will focus on working closely with subcontractors who share Morgan Ashurst's strategic and cultural objectives by continually communicating and practising the company's cultural values, including its commitment to:

- The Continuous Business Improvement Strategy
- Safety Health and the Environment standards
- Corporate Social Responsibility
- Involving subcontractors earlier and more fully in scheme development
- Focusing on various sectors, frameworks and investment-related work.

9.4.3 Consolidation

The majority of Morgan Ashurst staff felt that they traded with too many subcontractors. This lack of consolidation and the resultant fragmentation leads to various problems including an inability to set a uniform standard and establish effective teams. Coupled with this are the problems associated with resource waste and inefficiency along with a failure to reap the benefits of long-term collaborative working. The fractured link between the tender phase and construction phase also attracted a lot of discussion; subcontractors supporting Morgan Ashurst's tendering efforts were not always rewarded with orders. Similarly, the need to practise shared accounting was deemed an important issue. Despite Morgan Ashurst being required to provide clients with accurate estimates of the final account, and having its own policy of 'day one final accounting' which constantly updates the final account throughout the project, the subcontractors often submitted their accounts on an *ad hoc* basis. In addition, these submissions often failed to follow any recognised timetable – a prerequisite to enable Morgan Ashurst to provide competent cost reporting to its clients.

The case study highlighted the following issues:

- A focus upon transactions involving a smaller number of relatively high performing subcontractors;
- Gaining greater benefits from market leverage by placing work with fewer subcontractors, providing higher levels of income for that smaller group of firms;
- Investment of time to identify subcontractors with whom it is possible to develop shared strategic and cultural objectives;
- Simplification of processes and reduction in the administrative burden of low value orders by appropriately matching resources to tasks;
- Operating a 'joined-up' business where estimating and construction functions have identical objectives and are no longer susceptible to a silo mentality.

9.4.4   Consistency

Consistency is inextricably linked to consolidation. It was commonly felt that reducing the number of subcontractors with whom the organisation traded would enable a much more consistent approach. Consistency impacted in other ways:

- Transferable teams – The importance of maintaining teams that can draw on previous experience of working together. In areas of the business where this same logic has been applied to the entire project team, including subcontractors, it was seen as a key contributory factor to overall project success.
- Extend knowledge management to focus on subcontractors – the importance of making maximum use of the knowledge that exists both inside and outside the organisation.

Many participants stated that where they experienced better consistency, both internally and externally, it led to reduced duplication and confusion whilst enabling better decision-making. It was strongly felt that a large geographically diverse business working with a multitude of subcontractors on various projects and various sectors, needs to place special emphasis on being consistent in the way it:

- Delivers high quality and service standards;
- Integrates the pre and post-contract functions;
- Measures performance and provides feedback;

Part B

- Manages regional variances – a point that specifically acknowledges the differences amongst some of the markets in which Morgan Ashurst operates;
- Evaluates subcontractor quotations utilising multi-attribute selection criteria;
- Administers contracts and manages knowledge.

9.4.5   Cost

Participants were keen to talk about cost-related issues and felt that many high profile best practice reports had ignored these issues. As cost is an integral component of value, many participants believed it should not be sidelined in the development of the MASC. On some projects discussions between main contractors and subcontractors were limited to bilateral decision-making on the basis of lowest capital cost selection criteria. Cost growth during the subcontract programme was also a problem. Despite accepting substantial amounts of transferred risk from its clients, and seeking to manage that risk in a balanced way, the main contractor often experienced substantial non-recoverable cost increases from its subcontractor supply chain.

High subcontract procurement processing time, from enquiry stage through to contract completion and beyond, led to significant costs in some areas of the business. In addition, the time required to process subcontract orders by quantity surveyors diluted the ability of the main contractor to focus on its priority of generating value and instigating effective delivery and risk management strategies.

Participants pointed to their experience of the wider industry where many still operate in a paradigm where cost is simply regarded in terms of reducing one party's costs to increase another's profits. Whilst this undoubtedly holds true in certain instances, participants pointed to tangible examples where working towards the same goal had allowed all parties to reduce costs whilst protecting, and indeed increasing, profits through an emphasis on the more sophisticated concept of value. At the time of the study, Morgan Ashurst had invested in new information and communication technology systems with the aim of increasing cost transparency across the business. It was felt that cost was an area where further benefits could be made with relatively little investment. For example, analysing entry and exit costs (the difference between what a subcontract package of works is *expected* to cost and what it *actually* costs) and linking this information to future selection processes enables more intelligent subcontractor selection decisions to be made. Developing the focus on cost led Morgan Ashurst to focus on the following issues in its supply chain vision:

- Increasing the reliability of cost certainty and effectiveness in reducing costs, whilst generating value through closer working, shared savings, rebates and properly prepared objective cost analysis and comparisons.

- Analyse trade-specific proportional cost allocations and replace assumptions with accurate cost information to increase the effectiveness of decision-making.
- Identify unnecessary waste, such as over-specification, to reduce costs.
- Maintain the element of competition at both the pre- and post-contract phases of projects without undermining the principles of the SCM philosophy and the need to achieve continuous business improvement.

9.5   Conclusion

SCM has been a feature of several best practice reports (Latham, 1994 and Egan, 1998, for example) which sought to improve the effectiveness of the UK construction industry. The majority of these reports have a client-centric focus, which is surprising when one considers that few clients have ability to develop long-term partnered SCM. The distinction between organisational and project supply chains illustrates the important role that main contractors have in developing a managed supply chain. However, research has shown that contractor-subcontractor relationships are often less than harmonious. Indeed, Cox (2004a) and Cox *et al.* (2007) believe that all buyer-seller relationships are problematic owing to their essentially contested nature. It was argued that there is a real lack of practical advice for those main contractors wishing to develop their approach to SCM.

This action research case study sought to bridge the gap in the literature by focusing on a main contractor developing its own approach to SCM. The study has shed light on an area that has suffered from a lack of practical advice for practitioners seeking to implement their own SCM approach. It highlighted the key steps taken to develop a major contractor's approach and showed how consultation with internal and external staff was welcomed by those consulted, providing significant benefits in terms of gaining their commitment to the eventual change.

Action research was used in the case study covered in this chapter to inform future practice by focusing on the following issues:

- Relationships
- Culture
- Consolidation
- Consistency
- Cost.

The importance of relationships stood out as the single most important factor, impacting as it did on all the other themes. Participants could cite examples where long term relationships between contractor and subcontractors, allied to a shared concern for each others' interests, led to tangible examples benefits for all parties. The main contractor's ability to channel client demand and create an environment suitable for long-term collaborative subcontractor exchange transactions was made clear. The argument against win–win outcomes was challenged in favour of a more pragmatic notion of the client, subcontractor and main contractor 'winning' through

Part B

working together and sharing the same goals. Indeed, the adoption of a modified win–win–win triple concern model as a central thrust in its relational strategy indicates how important this issue was considered.

The MASC has undergone further development over and above that presented here and the results from these later stages are planned for future dissemination to enable other main contractors to learn from Morgan Ashurst's experiences.

References

Blake, R. and Mouton, J. (1964) *The Managerial Grid*. Gulf: Houston.

Briscoe, G., Dainty, A.R.J. and Millett, S. (2001) Construction supply chain partnerships: skills, knowledge and attitudinal requirements. *European Journal of Purchasing and Supply Management*, 20(7), 497–505.

Briscoe, G.H., Dainty, A.R.J., Millett, S.J. and Neale, R.H. (2004) Client-led strategies for construction supply chain improvement. *Construction Management and Economics*, 22, 193–201.

Briscoe, G.H. and Dainty, A.R.J. (2005) Construction supply chain integration: an elusive goal? *Supply Chain Management: An International Journal*, 10(4), 319–326.

Bullington, K. E. and Bullington, S.F. (2005) Stronger supply chain relationships: learning from research on strong families. *Supply Chain Management; An International Journal*, 10(3), 192–197.

Christopher, M. (1998) *Logistics and Supply Chain Management: Strategies for Reducing Cost and Improving Service*. 2nd Ed. Financial Times and Pitman Publishing: London.

Coghlan, D. and Brannick, T. (2004) *Doing Action Research in Your Own Organization* (2nd ed.). Sage Publications, London.

Construction Best Practice programme (CBPP) (2003) *Supply Chain Management*, CBPP: Garston.

Cox, A. (2004a) Win–Win? *The Paradox of Value and Interests in Business Relationships*. Earlsgate Press: Stratford-upon-Avon.

Cox, A. (2004b) Business relationship alignment: on the commensurability of value capture and mutuality in buyer and supplier exchange. *Supply Chain Management: An International Journal*, 9(5), 410–420.

Cox, A. and Ireland, P. (2002) Managing construction supply chains: a common sense approach. *Engineering, Construction and Architectural Management*, 9(5), 409–418.

Cox, A. Sanderson, J. and Watson, G. (2000) *Power Regimes: Mapping the DNA of Business and Supply Chain Relationships*. Earlsgate Press: Stratford-upon-Avon.

Cox, A. and Thompson, I. (1998) *Contracting for Business Success*. Thomas Telford: London.

Cox, A., and Townsend, M. (1998) *Strategic Procurement in Construction*. Thomas Telford: London.

Cox, A., Watson, G. Lonsdale, C. and Sanderson, J. (2004) Managing appropriately in power regimes: relationship and performance management in 12 supply chain cases. *Supply Chain Management: An International Journal*, 9(5), 357–371.

Cox, A., Ireland, P. and Townsend, M. (2007) *Managing in Construction Supply Chain and Markets*. Thomas Telford: London.

Dainty, A.J., Briscoe., G.H. and Millett, S.J. (2001) Subcontractor perspectives on supply chain alliances. *Construction Management and Economics*, 19, 841–848.

Department of Environment Transport and the Regions (DETR) (1998) *Rethinking Construction*, Report of the Construction Task Force, Department of Environment Transport and the Regions: London.

[Egan report] DETR (1998) *Rethinking Construction: The Report of the Construction Task Force to the Deputy Prime Minister, John Prescott, on the Scope for Improving the Quality and Efficiencies of UK Construction*, DETR: London.

Drayer, R.W. (1999) Proctor and Gamble's streamlined logistic initiative. *Supply Chain Management Review*. Vol. 3, Nr. 3, pp. 32–43.

Glaser, B. (1992) *Basics of Grounded Theory Analysis: Emergence vs. Forcing*. Sociology Press: Mill Valley, CA.

Fisher, N. and Morledge, R. (2002) Supply Chain Management. *In* Kelly, R.J., Morledge, R., and Wilkinson, S. (Eds). *Best Value in Construction*, pp. 201–221. Oxford: Blackwell Science.

Glaser, B.G. and Strauss, A.L. (1967) *The Discovery of Grounded Theory: Strategies for Qualitative Research*. Aldine, Chicago.

Green, S. (1999) The Missing Arguments of Lean Construction, *Construction Management and Economics*, 17, 13–137.

Green, S.D., Fernie, S. and Weller, S. (2005) Making sense of supply chain management: a comparative study of aerospace and construction. *Construction Management and Economics*, 23, 579–593.

Holti, R., Nicolini, D. and Smalley, M. (2000) *Building Down Barriers: The Handbook of Supply Chain Management – The Essentials*, CIRIA, London.

Hoole, R. (2005) Five ways to simplify your supply chain. *Supply Chain Management: An international Journal*. 10(1), 3–66.

Ireland, P. (2004) Managing appropriately in construction power regimes: understanding the impact of regularity in the project environment. *Supply Chain Management: An International Journal*, 9(5), 372–382.

Lambert, D.M., Stock, J.R. and Ellram, L.M. (1998) *Fundamentals of Logistics Management*. Boston: Mc Graw-Hill: Irwing.

Latham, M. (1994) *Constructing the Team*. HMSO: London.

Male, S.P. (2002) Supply Chain Management in Construction. *In* Smith, N.J. (Ed) *Engineering Project Management* 2nd Ed, pp. 264–289. Oxford: Blackwell Science.

Male, S.P. and Mitrovic, D. (2005) The Project Value Chain: Models for Procuring Supply Chains in Construction. *QUT Research Week Conference, 4th–8th July, Brisbane, Australia: RICS*.

McHugh, M., Humphreys, P. and McIvor, R. (2003) Buyer-Supplier relationships and organizational health. *Journal of Supply Chain Management*, Vol. 39, Nr. 2, pp. 15–25.

Mouritsen, J., Skjott-Larsen, T. and Kotzab, H. (2003) Exploring the contours of supply chain management. *Integrated Manufacturing Systems*, 14(8), 686–695.

Muya, M., Price, A.D.F. and Thorpe, A. (1999) Contractors Supplier Management, in Bowen, P., and Hindle, R. (Eds) *Proceedings of CIB W55/65 Joint Trienial Symposium, Customer Satisfaction: A Focus for Research and Practice in Construction, 5–10 September, Cape Town*.

Poirier, C. and Houser, W. (1993) *Business Partnering for Continuous Improvement*. Berrett: Kohler, CA.

Reason, P. and Bradbury, H. (Eds.) (2001) *Handbook of Action Research: Participative inquiry and practice*. Sage: London.

Strategic Forum (2002) *Rethinking Construction: Accelerating Change*, consultation paper, Strategic Forum for Construction: London.

Strauss, A. and Corbin, J. (1998) *Basics of Qualitative Research: Techniques and procedures for developing grounded theory*, 2nd Edn. London: Sage.

Yin, R.K. (1994) *Case Study Research: Design and Methods*, 2nd Edition. Newbury Park, CA: Sage.

Part B

10

Franchising the Supply Chain

Hedley Smyth

Supply chain management (SCM) provides a means to increase control over companies that are not owned by the client or first-tier supplier. This view involves intervention into the affairs of another firm without the responsibilities of ownership. Intervention and control can take many forms. The most common form of control involves little or no direct intervention. Control is indirect control in the form of a supply contract. This lack of intervention links supplies, but cannot be said to be SCM. Something more or something else is required to control or manage the supply chain from the next tier and certainly if more than one tier in the chain is to be managed. Partnering and SCM agreements are one way of achieving direct intervention, although in construction it has been argued that these represent means simply to drive prices down rather than drive added value up (e.g. Green, 2006).

Another way to manage the supply chain is to use franchising principles. In essence, franchising has the same objective as SCM – to intervene and exert some control of supply without having ownership of the supply process. This is achieved by granting a licence to an operator/subcontractor to provide a service and deliver a product in detailed terms set out by the customer that go beyond the requirements of a typical contract. Franchising increases the extent of customer/client intervention and indeed increases the presence of the franchisor in the market and, therefore, has greater market impact for the franchisor than achieved through SCM. In the same way that SCM cannot be transferred uncritically from other sectors to construction in order to have the same level of effect (e.g. Smyth, 2005), the same holds for franchising the supply chain. This chapter explores how franchising principles can be applied in construction supply chains and draws attention to the ways in which it has been applied, drawing lessons for the potential and application to practice on a broader scale in the future. The chapter starts with an overview on the growth and application of franchising in general and relates this to construction supply chains for each principle. The chapter subsequently focuses specifically upon franchising for construction at a detailed level, conceptually supporting the analysis with empirical examples.

10.1   Towards Controlling the Supply Chain

10.1.1   From diversification to core business

During the 1970s and 1980s, firms in many sectors focused upon 'core business', thus 'outsourcing' activities which were not perceived as core. Whilst construction had been a leader in subcontracting, it was the defence industry that provided an important model for subcontracting or outsourcing work at this time. Defence projects had become more demanding in size and technical complexity. Primary providers could do less in-house and simply required a wider variety of suppliers to meet growing demands. Another trend was the increasing costs of internal administration of large diversified companies and the internal inefficiencies that led to in-house provision being less competitive than outsourcing – a 'make' or 'buy' question in transaction cost terms (e.g. Williamson, 1985). In addition, resources are limited and it made greatest competitive sense to concentrate resources in core areas of strengths – allocation of resources into the core business within the resource based view (e.g. Barney, 2001).

In parallel, stock market investors were applying portfolio management models. This enables investors to manage a wide diversity of investments and spread risks, without having to invest in diversified companies who undertake the risk-spreading on behalf of investors by default. Pressure was therefore exerted by investors on producers to focus upon their core business and sell off non-core activities – resulting in enforced 'buy' decisions in terms of formerly vertically integrated diversified companies.

10.1.2   From core business towards managing supply chains

One consequence of focusing upon core business is that change in one area of product design, manufacturing process or delivery can have knock on effects on suppliers. Different ownership can act as a barrier to making changes as decision-making lies outside the remit of management. Intervention in the form of SCM is one solution, whereby firms collaborate in order to make changes to the specification of products and processes to achieve more efficient and effective solutions for the marketplace.

Such changes can be motivated in two ways. SCM can be used to coordinate processes that increase profit through efficiencies – reducing costs through collaboration on product design, lean production and just-in-time deliveries for example. These are production and procurement based drivers for SCM. Construction has primarily focused upon a procurement orientation of SCM in practice (e.g. Smyth, 2005), largely driven by clients and through industry reports (e.g. Egan, 1998).

The extent to which production and procurement efficiencies can be handed on to customers, in part or in total, depends upon market demand in general terms, and the relative market power of customers and key players along the chain. In construction, the customer or client tends to have considerable power or *leverage* (Cox and Ireland, 2006) even in all but the most buoyant markets as the supply side has, in past at least, been fragmented and largely competing upon price (e.g. Smyth, 2000; 2006). Therefore the benefits of SCM tend to accrue disproportionately to clients (see Chapter Five by and Skitmore and Smyth).

Firms along the supply chain can collaborate for a second reason – a motivation to be customer or client orientated. In practice 'adding value' frequently includes better value for money by reducing costs to the customer, which can arise from the production-orientation (e.g. a lean approach or value management) and procurement-orientation (driving costs down through procurement measures). Added value, in its literal or pure sense, provides an increase in the content of the product or service than otherwise would be the case. It may be for the same cost to the customer, for a proportional increase in costs or a premium cost because a premium profit margin is secured on the added portion of work. How this works out in practice again depends upon the market power of the procurer at each stage along the chain, but also depends upon the level of benefit the customer/client derives from the added value.

Construction has not traditionally been client-orientated. A procurement driver from clients would logically be met with a marketing response from contractors, yet contractors have tended to drive the procurement initiative, such as the request for lower client costs or higher added value, along the chain (Smyth, 2005), typically to the next tier and occasionally the tier below that (Olayinka and Smyth, 2007; Mason, 2008). The consequence is that contractors relinquish direct responsibility to add value.

Changes to product and process specification can range to marginal changes with minimal cost to large changes requiring substantial investment. In manufacturing, investment expenditure may be a substantial up-front cost for supply chain members, yet is spread over multiple units of output making unit cost increases relatively small in mass production and even in small batch production. In construction, main contractors produce relatively small amounts using in-house operatives, acting as managers and entrepreneurs in the market place. The burden of changes to product and process therefore inevitably lies with *subcontractors*. They are consequentially at a distance for client expectations and requirements, having and being given little incentive to become more involved except via the client. In addition, every project is different and so any design-cum-product investment is one-off and must be recouped within a single project. And yet, contractors and subcontractors alike are required to manage projects and the management processes do have repetition in terms of applying bodies of knowledge (e.g. PMI, 2004; APM, 2006), management approaches such as Prince2, plus project management tools and techniques. There is therefore, potential for process innovation on the part of subcontractors and contractors. There can also be similarities in management across projects of one contractor with potential for consistency and continuity of approach between them (Smyth, 2000; Pryke and Smyth, 2006; Wilkinson, 2006). There is clearly scope for applying SCM to add *service value* (rather

Part B

than product value), which contractors have neglected according to clients (Pratt, 1999; Mason, 2008). These service issues are more akin to retailing, a field where franchising has found favour, than mainstream manufacturing *per se*.

10.1.3   From core business to franchising

Firms that are responsible for managing the point of sale – the exchange and related transaction costs – are typically retailers. They have benefited from technologies of inventory control and just-in-time deliveries, which are linked to the production and procurement orientation of manufacturing SCM. These efficiency gains have permitted them to minimise stock areas, maximise display areas and increase the variety of stock on offer. An example is the way in which lines of fashion stock in retail outlets change weekly, rather than maintaining the same garment range over an entire season. These features, plus the production of clothes in low cost countries have combined to stimulate demand and present opportunities for retailers (as well as other organisations) to offer their goods locally through multiple outlets. The cost of expanding is high; the costs of managing new outlets in a consistent way to maintain quality and brand image, are also high.

Therefore, many retailers reconceived their core business as managing the business processes, rather than always managing the retail outlets *per se*. Retailers such as French Connection, renaming outlets as fcuk, have a mix of owned and franchised outlets. Many food and drinks retailers franchise all outlets where greater standardisation is achievable. McDonalds is the archetype, but many restaurants, fast food, coffee companies extensively apply franchising. These businesses standardise their management processes. In the case of McDonalds, the configuration of the outlets and the technologies within them create routines; routines for management and customer care have refined by management and employees learn through training (basic cognitive learning coupled with administrative and social skills) and on the job learning (basic experiential and psycho-motive learning) and perhaps the use of communities of practice (e.g. Wenger, 1998; Robbins, 2003; cf. Nelson and Winter, 1982).

Defining the parameters of franchising will be developed fully in the next main section. At this stage the salient points are that franchising introduces another link in the chain. The owner invites other businesses to enter the market and operate outlets on their behalf using their technologies, management systems and procedures, their products/suppliers and brand. The franchisee introduces the operating capital and hence brings additional resources to expand the outlets. The franchisor takes a licence fee and typically a share of the profits. The franchisee carries most of the risk.

In construction, contractors introduce another link when they subcontract. The subcontractors undertake work on site. Contractors have found that insufficient control can be achieved from Head or Regional Offices, tending to use site-based contracts managers and contract directors. This decentralisation keeps risk low by minimising investment, whilst operational risk is transferred to the subcontractors. This is a procurement-orientation from the contractor perspective and inhibits a client-orientation for adding service value. Franchising a management approach to subcontractors may carry some up-front investment for contractors and some administrative costs to ensure quality control, yet offers a lower cost approach that can focus upon the first tier of the chain. This is an alternative, and possibly a lower cost approach than applying SCM principles for service continuity and consistency on all projects.

In other words, contractors have shied away from incurring the marketing and procurement management costs of SCM, restricting activity to the unsophisticated management of procurement by simply passing the procurement driver along the chain to squeeze costs and 'add value' in an uncontrolled way. Franchising the supply chain offers an alternative. This chapter continues by exploring the scope for franchising in greater detail generally and for construction. It is not normative in the sense of saying that this is what contractors should do. It is also recognised that possible take up would be confined to a market segment of contractors and suppliers trying to differentiate their services from other providers. It is presented as a challenge to the construction industry in particular. In that sense the challenge is a test of how progressive the industry is, and is prepared to be, evaluation coming further down the line when management strategy and activities can be evaluated reactively. It also provides part of a continuing critique of the client-driven procurement agendas which have failed to adequately articulate how concepts, including partnering, SCM and lean production can be translated into the sector in ways that address the requirements of all parties. Clients and reports associated with client-focused issues (e.g. Egan, 1998) construe progress and continuous improvement along very narrow lines, failing to address the diversity of approaches which contractors could take and hence the differentiation of service delivery to fit diverse clients – a market and client-orientated approach (Smyth, 2000, 2004, 2006; Smyth and Edkins, 2007; Smyth and Fitch, 2007).

10.2   Conceptualising Franchising

10.2.1   Defining and conceptualising of franchising

Franchising involves granting a licence to an operator/subcontractor to provide a service and deliver a product, typically for a period of some years. The way in which the service is delivered is specified in considerable detail, beyond the requirements of a typical contract and usually in line with the way in which the franchisor would manage the process if they were doing in-house. Hadfield (1990) stated that franchising provides a hybrid between employment and independent contracting, whereby both the franchisor and franchisee contribute to production and service delivery.

Franchising aims to achieve a series of possible objectives for the franchisor:

- Access to resource injection for rapid expansion, typically geographical, but also penetration of existing markets;
- Spreading of risks of expansion (e.g. Lafontaine, 1992);
- Income from licensing brand name and frequently including a profit share (e.g. Rubin, 1978);
- Maintenance of product and/or service quality.

Franchising aims to achieve a series of possible objectives for the franchisee:

- Achieving rapid economies of scale (e.g. Caves and Murphy, 1976);
- Reducing start-up risks for a new business, and marketing and sales costs for an existing business;
- Access to customer base through brand reputation (e.g. Lafontaine, 1992);
- A tried and tested product/service offer.

There are trade offs and balances to achieve for successful franchising. Part of the brand is for the franchisee to standardise key technology and key operational-cum-service components. The consequence is that the training, management start-up (new business) or adaptation (existing business), ongoing support and expert advice, quality monitoring and feedback are all part of the franchising package, the costs of which have to be divided up between the parties. The division must strike a balance between encouraging entry into the franchise market and costs sufficient to put off 'free rider' entry. Similarly, the distribution of returns must be sufficient reward for both parties. Some franchisees are specialist companies that take on multiple businesses, for example signing-up for a whole region of coffee shop outlets. These specialists have experience in negotiating balanced franchise packages.

Franchisors need to be cautious about being over-ambitious in securing income to feed expansion in the early stages. For the franchisee the duration of a franchise agreement is typically critical, not only to encourage entry, but also to provide sufficient time to pay for historic costs such as investment in set up, sunk costs for training, and investment for their own business expansion in the future. The maintenance of healthy franchise relationships typically involves a degree of transparency concerning the way that all franchisees are secured and managed, periodic review meetings with all franchisees to share new developments and experiences, regular franchisor-franchisee meetings; field visits from expert advisors to support franchisees, which is kept as a separate function from activities to monitor quality (JBR Hellas, 2001). Grievance procedures usually form part of a franchise agreement to aid relationship maintenance.

Categories of franchise development and business models have been identified, including investment franchises; management franchises; executive; retail; distribution; and 'man and van' franchises. The balance and combination of factors to induce successful franchising operations depend upon the context found in the respective categories. Construction presents a specific context.

10.2.2   Towards franchising in construction

The potential for construction franchising falls into two of the identified categories, namely management franchises and the man and van franchises. The management category and the man and van category provide the two options. Management franchising is appropriate for the first tier of the chain. The main contractor can apply the following options in this category:

- Select specialist contractors based upon existing service quality, possibly narrowing the number of suppliers to a maximum of two or three;
- Extend their own brand by requiring the subcontractor-franchisees to use the main contractor brand on site, vehicles, clothing and documentation, this providing an alternative to increasing direct labour to improve management control and quality, and also providing the main contractor with specialist capability when bidding in particular market segments or for particular types of work;
- Specify technologies and the management of the technologies where appropriate, especially for certain specialist subcontract work or work that is critical to the market of the main contractor, and potentially in the long term working with the subcontractor-franchisee to develop technologies for specialist work;
- Specify output at a higher level of quality, perhaps working with the subcontractor-franchisee;
- Specify management systems and procedures by which all work will be conducted, perhaps linking to just-in-time methods;
- Possibly specify codes of behaviour on site and in the long term at team meetings;
- Requiring the first tier subcontractor-franchisee to ensure that any further subcontractors they appoint conform to the franchise principles.

The investment and commitment from the main contractor would involve:

- Consideration of existing management systems and procedures, plus any codes of behaviour to ensure fairness and consistency;
- Training provision, and quality monitoring systems for subcontractor-franchisee output and organisational behaviour;
- Availability of dedicated expertise advisors for day-to-day issues and technical/management staff for developing collaborative working.

Achieving a balance between the main contractor requirements as franchisor and the subcontractor-franchisee requirements is context specific. A range of options is available, which is explored below from the subcontractor viewpoint. The subcontractor is likely to be an existing company in the management franchising category, unless a direct labour division or subsidiary is being hived off as an independent operation. In this position a subcontractor requires a balanced approach.

Repeat business is a major incentive and competing or being allocated work against a maximum of two other suppliers would provide reasonable certainty in most markets. It is unlikely that subcontractors will commit to exclusively to one customer/client, and therefore will pursue other work in traditional subcontracting market. It is important that the main contractor controls quality, ensuring that operatives are not transferred from one site to a franchised project without the correct branding applied in a fashion that adds to the image of the main contractor. The same holds for techno-logical and management specifications.

The commitment cost of subcontractors to becoming franchisees is con-siderable, adding to overheads and investment. A licence fee and standard franchisee 'profit share' or 'rent payable' may not therefore be affordable in the construction market. There may be a nominal 'turnstile charge' to gain access to the market in order to put off the uncommitted opportunistic and free rider firm, but an annual fee may be prohibitive. An enhanced gain share approach on individual projects provides profit incentives, penalties (the pain aspect) being incurred for non-conformance to franchise princi-ples. An escrow account could also be opened to invest penalty charges in developing performance and relationships, as applied in areas of the IT sector (Birt, 2003).

Support from the main contractor – training, expertise advice and exper-tise collaboration, plus forums for sharing practices and experiences – is vital for the franchisee. This provides a point of departure for main contrac-tors, whereby they need to move towards programme management to sup-plement their management of projects (Smyth and Pryke, 2008) in order to develop the necessary support.

One of the problems at the outset with partnering was a lack investment from contractors because clients were reluctant to specify commitment, hence duration across projects in the absence of framework agreements. The contractual commitment envisaged originally by Latham (1994) was never fully realised due to the potential for claims arising for breach of contract, where long term assurances of workload are involved. Contractors also tended to change the rules, for example, the number of selected partners. In franchising, the franchisees are carrying greater risks and some certainty is needed. A reasonable franchise period is needed to defer costs and incur a return on the additional investment made. A degree of competition for jobs – maximum three bidders without any risk of switching or bidding against traditional subcontractors – rather than the captured geographical markets in retail – would overcome pricing issues and could be supported at regular face-to-face meetings by discussion of how rates are built for certain types of work using open book principles to lend confidence to pricing of bids from franchisees. Main contractors may need to give minimum thresholds of contract value that franchisees will be asked to bid for over the contract period. This could be based upon a percentage of contract value that the main contractor bids for in order to overcome fluctuations in overall market demand.

A main contractor may develop franchising by first applying the principles to a simple service, for example painting and decorating. Where large painting contracts are involved, for example for redecoration of social housing accommodation, the following principles could be applied:

- Applying main contractor brand to vehicles, clothing and documentation;
- Specify quality of output, perhaps brand of paint;
- Specify procedures by which all work will be conducted, including working periods within the programme;
- Specify codes of behaviour, which make include guidance on politeness to occupants, no swearing, nor use of electronic equipment on site, and no use of mobile phones within an apartment;
- Health and Safety and Customer-Care training for operatives;
- Establishment and maintenance of places on Local Authority Lists of Approved Contractors, in a way that is out of the reach of small businesses due the financial capacity requirements imposed as prequalification for such lists.

Refinement and extension of the principles could then be developed for more complex and specialist subcontracting.

Painting and decorating requires management principles, yet shares some ground with the man and van category of franchising. Man and van franchising is appropriate for the last tier of the chain or for simple franchising operations. The subcontractor-franchisee benefits in the following ways in this category:

- Maintain independence and self-employed status;
- Receive training for market entry and support in the market;
- Secure access to work through a larger organisation;
- Secure license to operate within a geographical area, particularly where specialist technology is required or where FM and maintenance work is conducted;
- Access to specialist Health and Safety advice and documentation;
- Access to a wide geographical coverage of Local Authority and other Public Sector Lists of Approved Contractors, backed by a central administrative function to maintain such places.

This model can enable major contractors to rapidly expand into the largest area of construction work, the repair and maintenance market, without carrying the management costs and risks for local operations. In the man and van franchising model an annual licensing fee and profit share is applicable. An initial 'turnstile charge' above the license fee is inappropriate, because this would act as a barrier to market entry at this level, but charges for training might be made at entry point or recouped over subsequent years (perhaps using the principles of the student loan system).

Part B

10.2.3   Current franchising in construction

Watson and Kirby (2000) considered franchising in construction. They argued that the risk of managing franchisees is high, yet this is a function of unfamiliarity in the sector and would be overcome through experience. They also argued that the risk profile is attractive to franchisees, particularly within the man and van category identified in this chapter. Watson and Kirby (2000) conducted interviews with franchisees in the construction sector, finding that there were few operational problems in practice. This suggests the model, especially in the man and van category could be extended. However, the analysis of franchising principles applied to construction also showed that management franchising would be optimally pioneered on simple operations, which have parallels with man and van franchising model. They can then transition to more specialist activities as experience is gained.

Using the UK as an example, franchising in construction shows there are a range of franchisees – see Table 10.1.

Part B

Table 10.1   Examples of UK franchising

Company	Type service	Primary features	Category
Allied Preservation	Renovation and environmental services	Geographical	Man and van model
Aspray Building Services	Project management	Geographical	Management model
Aqua-Lec	Plumbing repairs and equipment maintenance	Geographical	Man and van model
Calbarrie	Entertainment electrical Contractor	Geographical	Man and van model
Concept Building Solutions (UK)	Insurance repair services	Geographical	Man and van model
Dyno-Rod	Plumbing services franchise Drainage services franchise	Geographical Specialist technology	Retail model and man and van model
Elec Local	Electrical services	Geographical	Man and van model
Homeserve	Insurance repair services	Geographical	Man and van model
Pave It UK	Drives, pavements and patio Construction	Geographical	Man and van model
Plumb Local	Installations	Geographical	Man and van model
The Power Service	Utility repair and maintenance	Geographical	Man and van model
Sumo	Supplier of ground penetrating radar services	Specialist technology and skilled operatives	Man and van model

Source: Internet search using http://www.startups.co.uk

Whilst the dominance of the man and van franchising model is apparent, there are large contracting organisations behind some of these cases, for example Dyno-Rod is part of the Centrica group and has resource backing from British Gas. This operation shows some features of the retail franchise model as well as the man and van model which is more appropriate in mainstream contracting. The majority of the services are located in the repair and maintenance and facilities management markets. However, Sumo offers technology based services and Pave It UK offers new construction services, providing pointers towards the management franchising model.

10.3   Conclusion

This chapter has explored the concept of franchising. It has considered the conceptual scope of application to the supply chain and has briefly reviewed the current state of franchising in UK construction markets. Two franchising models have been identified as conceptually most applicable: (i) management franchising, and (ii) man and van franchising. Empirical evidence shows prevalence for the man and van model, yet the conceptual analysis has shown that there is some overlap whereby large main contractors have scope to explore and gain experience following man and van procedures, transitioning towards application to more demanding and complex specialist services over time. In one sense, the application of franchising has more to offer the more technically complex trades (like mechanical and electrical services) than less complex trades like painting.

The chapter also presents a critique of the predominance of procurement models. It offers a broader approach and thus is also an implicit critique reform movement based upon relational contracting to secure continuous improvement. There have always been other conceptual options for developing improvement. Franchising, applied to the supply chain, provides one alternative means for controlling product and especially service value for main contractors and clients.

Part B

References

APM (2006) *Project Management Body of Knowledge*, Morris, P.W.G. (ed). Association for Project Management, Buckinghamshire.

Barney, J.B. (2001) *Gaining and Sustaining Competitive Advantage*. Prentice Hall, Upper Saddle River.

Birt, D.N. (2003) Supply Chain Challenges: building relationships. *Harvard Business Review*, July, 64–73.

Caves, R.E. and Murphy, W.F. (1976) Franchising: firms, markets and intangible assets, *Southern Economic Journal*, 42(4), 572–586.

Cox, A, and Ireland, P. (2006) Relationship Management Theories and Tools in Project Procurement, *The Management of Complex Projects: a relationship approach*. Pryke, S.D. and Smyth, H.J. (eds.), Blackwell, Oxford.

Egan, Sir John (1998) *Rethinking Construction*, HMSO, London.

Green, S.D. (2006) Discourse and fashion in supply chain management, *The Management of Complex Projects: a relationship approach*. Pryke, S.D. and Smyth, H.J. (eds.). Blackwell, Oxford.

Hadfield, G.K. (1990) Problematic relations, franchising and the law of incomplete contracts, *Stanford Law Review*, 42(4), 927–992.

JBR Hellas (2001) In thought of Franchise Requirements for a Proper Cooperation, Franchisor Franchisee Relationship, JBR Hellas, Business Consultants.

Lafontaine, F. (1992) Agency theory and franchising: some empirical results, *The RAND Journal of Economics*, 23(2), 263–283.

Mason, J. (2008) Specialist contractors and partnering, *Collaborative Relationships in Construction*, Pryke, S.D. and Smyth, H.J. (eds.), Blackwell, Oxford.

Nelson, R.R. and Winter, S.G. (1982) *An Evolutionary Theory of Economic Change*, Harvard University Press, Boston.

Olayinka, R. and Smyth, H.J. (2007) Analysis of types of Continuous Improvement: demonstration projects of the Egan and post-Egan agenda, *Proceedings of RICS Cobra 2007*, 6–7 September, Georgia Institute of Technology, Atlanta.

PMI (2004) *A Guide to the Project Management Body of Knowledge*, 3rd edition, Project Management Institute, Newton Square, PA.

Pratt, J. (1999) Understanding clients – the Egan imperative, *Proceedings of the 4th National Construction Marketing Conference*. July, Centre for Construction Marketing in association with CIMCIG, Oxford Brookes University, Oxford.

Pryke, S.D. and Smyth, H.J. (2006) *The Management of Projects: a relationship approach*, Blackwell, Oxford.

Robbins, S.P. (2003) *Organisational Behaviour*. Prentice Hall, New Jersey.

Rubin, P.H. (1978) The theory of the firm and structure of the franchise contract, *Journal of Law and Economics*, 21(1), 223–233.

Smyth, H.J. (2000) *Marketing and Selling Construction Services*. Blackwell, Oxford.

Smyth, H.J. (2004) Competencies for improving construction performance: theories and practice for developing capacity, *The International Journal of Construction Management*, 4(1), 41–56.

Smyth, H.J. (2005) Procurement push and marketing pull in SCM: the conceptual contribution of relationship marketing as a driver in project financial performance, *Journal of Financial Management of Property and Construction*, 10(1), 33–44.

Smyth, H.J. (2006) Competition, *Commercial Management of Projects: Defining the Discipline*, Lowe, D. and Leiringer, R. (eds.). Blackwell, Oxford.

Smyth, H.J. and Edkins, A.J. (2007) Relationship management in the management of PFI/PPP projects in the UK, *International Journal of Project Management*, 25(3), 232–240.

Smyth, H.J. and Fitch, T. (2007) Relationship management: a case study of key account management in a large contractor, Paper presented at *CME25: Construction Management and Economics: past, present and future*, 16–18 July, University of Reading, Reading.

Smyth, H.J. and Pryke, S.D. (2008) *Collaborative Relationships in Construction: developing frameworks and networks*, Blackwell, Oxford.

Watson, A. and Kirby, D.A. (2000) Explanations of the decision to franchise in a non-traditional franchise sector: the case of the UK construction industry, *Journal of Small Business and Enterprise Development*, 7(4), 343–351.

Wenger, E. (1998) *Communities of Practice: learning, meaning and identity*. Cambridge University Press, Cambridge.

Wilkinson, S. (2006) Client handling models for continuity of service, *Management of Complex Projects: a relationship approach*. Pryke, S.D. and Smyth, H.J. (eds.), Blackwell, Oxford.

Williamson, O.E. (1985) *The Economic Institutions of Capitalism*, Free Press, New York.

Part B

Conclusion

Stephen Pryke

The purpose of this conclusion is to summarise the key points from the text, to ask how these points contribute towards the debate about SCM in construction, and to identify how the knowledge and information might be applied in the construction industry. In addition, there is the inevitable speculation about what might have been dealt with differently and what needs to be done in the future – any unfinished business. The main headings for this discussion are as follows:

- Origins, myths and motivations relating to supply chains and their management;
- A summary of the contribution that this book makes;
- A sense of what it means conceptually and in the application;
- Suggested next steps.

11.1.1   Origins, myths and motivations

Early post-Latham (1994) reform in UK construction placed a great deal of emphasis on dealing with adversarial relationships through relational approaches to contracting; those relationships within project coalitions that focused on problem-solving and continuous improvements. There was a range of approaches to structuring and managing these collaborative relationships. The public sector through Prime Contracting (Holti *et al.*, 2000) adopted the approach of managing the supply chain through an 'agent' – using major contractors, or other types of firm, to manage the supply chain on behalf of the client. Although the Tavistock Institute established some important principles for the application of SCM in any sector, the Prime Contracting experiment was deemed to have been less than a resounding success for the Defence Estates organisation and procurement and project management ideas moved on for the UK public sector. Meanwhile ex-public sector British Airports Authority (BAA) adopted a client-led, essentially bureaucratic, approach to managing its supply chains. BAA made a major contribution to the development of SCM in UK construction, and it is fitting that a chapter in this book is devoted to their initiatives. Finally, Slough Estates adopted a non-bureaucratic, relationship-focused approach (Pryke and Smyth, 2006) to managing their supply chain. A review of the BAA chapter by Potts (**Chapter 8**) and that of Slough Estates by Rimmer (**Chapter 7**) is given below.

The introduction to the book deals with the quite specific references that the Egan Report (1998) makes to:

- Acquisition of new suppliers through *value* based sourcing;
- Organisation and management of the supply chain to maximise innovation, learning and efficiency;
- Supplier development and measurement of suppliers' performance;
- Managing workload to match capacity and to incentivise suppliers to improve performance;
- Capturing suppliers' innovations in components and systems.

It is hoped that this book has contributed some answers as to whether these aspirations have been delivered.

11.1.2   A summary of the contribution made

Parts A and B deals with concepts and application respectively. **Chapter 1**, comprised the introduction to the book. This first chapter provided some context for the material that followed including an overview of partnering in UK construction. The balance of the chapter provided an overview of the contents of the book.

Chapter 2 by Morledge, *et al.* dealt with the development of SCM from its origins in manufacturing. The importance of cultural change as a vehicle for the pursuit of the Latham (1994) and Egan (1998) agendas was identified and the authors wrestled with the problems associated with the implementation of SCM principles by one-off or low volume clients. Morledge *et al.* looked at a number of problems within UK construction that might arguably be addressed through the use of SCM. These are:

- Fragmentation
- Adversarial relationships
- Project uniqueness
- Separation of design and production
- Competitive tendering.

Morledge *et al.* identified the involvement of subcontractors in detailed and perhaps even conceptual design as an important benefit flowing from supply chain management. They concerned themselves with the difficulty of achieving what might be referred to as 'bottom up' design in a procurement process that appoints the typical specialist subcontractor at a stage when many important design decisions have had to be made by project actors that

Part B

will not be responsible for the construction of that particular element of the building. The use of clusters (Gray, 1996) embedded within a partnered supply chain managing approach was cited as a possible solution to the problem. Morledge *et al.* concluded by expressing some enthusiasm for SCM on the basis that construction is not a single industry at all; the diverse types of firms in a typical project coalition being best managed using a supply chain approach. They do, however, express a certain discomfort with the very term *supply chain management*, expressing a preference for *supply networks*. This is a theme that is also discussed in **Chapter 1,** where some attempt is made to rationalise these two conflicting positions.

Fellows dealt with culture in supply chains in **Chapter 3.** He reviewed the seminal work of Hofstede (1994) and Schein (1990), among others, before exploring the various ways in which culture might impact upon the construction supply chain. The idea that (*inter alia*) value is added as a given project component or service flows through the supply chain is complicated by the different perceptions of value held by each of the individuals that comprise a given project actor firm; these values can be business, technical and personal. The competition for resources and programme time between individuals and firms can cause a skewing of the values delivered to the client and stakeholders. Culture is not just about how we do things but also what, why, when and by whom (Schneider, 2000). Each of these aspects of human relationships has an impact on the extent and effectiveness of supply chains. Fellows cited Elmuthi and Kathawala (2001) who dealt with the main cultural issues associated with alliances and which Fellows suggested are equally applicable to supply chains. These limiting factors are:

- Clash of cultures;
- Lack of trust;
- Lack of clear goals and objectives;
- Lack of coordination between teams;
- Differing procedures and attitudes;
- Relational risk associated with self-interest focus.

These issues help us to understand the sources of difficulties faced in a range of collaborative relationships including construction supply chains.

Bresnen continued the theme of culture into **Chapter 4,** but rather than the focus of the chapter, culture became one of the issues affecting learning and innovation in supply chains. In this chapter on learning, knowledge and innovation, Bresnen explained how culture can stand in the way of learning and reflects upon the problem that construction has in relation to collaborative initiatives in supply chains frequently being isolated within the first tier of the supply chain. Bresnen (in common with Fellows in **Chapter 3** above) looked at leverage or power in supply chains (citing Cox, *et al.,* 2001). He explained that although leverage can be associated with commercial power and expertise, in the context of any given project-orientated transaction, there are also '. . . deeper underlying systems of rules and norms that govern interaction in an interorganisational setting and which, therefore, may effectively privilege one group of interests over another' (Cox *et al.,* 2001).

Bresnen's chapter looked at project-based organisations and the way in which they tend to inhibit innovation and the *cooperating to learn* function associated, potentially, with close supply chain relationships. As a result learning and innovation tend to be associated with the *local and particular*, very little finding its way to a broader, industry-based, audience. Bresnen feels that SCM is underdeveloped in construction and saw a need to develop SCM for the purposes of knowledge creation and transfer, innovation and learning. Bresnen concluded that there needs to be a move away from a project-based mindset towards a supply chain-based mindset – in this way there is the potential for more explorative learning to occur.

Skitmore and Smyth looked at marketing and pricing strategy in supply chains in **Chapter 5**. They argued that in order to achieve the effective addition of product and service value in construction supply chains, there needs to be some emphasis placed upon marketing and pricing strategies. This perhaps calls for a reappraisal in the way that SCM is understood; in particular, there is criticism of what is an essentially deterministic approach to the discussion and analysis of supply chains. The chapter looked at the familiar problem in construction that supply chains and SCM are perceived in a variety of ways within construction and this perception inevitably affects the way in which supply chains are construed and the form which management of these supply chains takes. The authors proposed organisationally driven relationship marketing as an alternative to the procurement-driven approach that tends to prevail in construction. Skitmore and Smyth's contribution involves pricing theory and relationship marketing; the predominance of the price-dominated version of the marketing mix in construction is brought into question and consequential cost cutting and value reduction. It is suggested that where there is a context in construction, involving routine risk minimisation coupled with transaction cost emphasis, and this simply creates a situation where costs are cut to achieve competitive status; value added may also be reduced and continuous improvement is unlikely to flourish. **Chapter 5** provides some unconventional and innovative ideas about construction supply chains and some illuminating thoughts about their operation and understanding. This chapter marks the end of the predominantly conceptual material.

The second part of the book dealt with the application of supply chain management in practice, within the UK construction industry, starting with risk. Part B contains two chapters that looked at supply chain management from the viewpoint of the client organisation. They adopted quite different approaches: Slough Estates with an essentially informal, relationship-based system; BAA with its structured, literally bureaucratic system. The chapter that followed (**Chapter 9**) focused upon the interests of contractors. Finally, **Chapter 10** provided a model for subcontractors to use, although whether it comprises SCM, or an alternative to SCM, is for the reader to decide.

Edkins opens Part B with a chapter that dealt with risk management and the supply chain in **Chapter 6**. He observed that if there is a distinct discipline of SCM it is difficult to isolate and identify, and as an area of academic study, particularly in relation to construction, it is still in its infancy. He took the position that supply chains evolve and are maintained according the principles of economics – they will be maintained only where the transaction cost of maintaining them is less than the benefit accrued by individual supply chain members from their involvement in a given supply chain. He put forward vertical integration as an alternative to supply chains (which are essentially a function of outsourcing – the decision to buy rather than make) and posited that the rise of the supply chain is largely a reflection of the dynamism needed to exploit rapidly changing markets or opportunities, but at the same time maintaining flexibility in the face of uncertainties in future workload. One of the interesting hypotheses offered within this chapter was that risk comes to rest in the supply chain at the position where leverage is dominant on the part of the *transferring out* project actor. In other words, whereas it would be optimum in risk management terms for risk to be managed by the actor within the project that is best placed through virtue of that actor's knowledge, experience and resources, frequently the reality is that the actor in the supply chain network that is least able to defend itself against other actors linked to it in some way, will eventually have to absorb and manage the risk somehow. All of this is in complete disregard for any financial compensation which may or may not be available to the *risk receiving* party. Edkins suggested that economic power is not particularly relevant to supply chain members and that the power or leverage exercised is supply chain specific and related to the power of the other firms within the supply chain.

Rimmer contribution to Part B provides discussion and analysis related to Slough Estates plc. The chapter (**Chapter 7**) relates to the hands-on supply chain managers in industry. Slough Estates had the foresight and courage to provide resources and the necessary expertise to identify, understand and manage collaboratively the supply chains used by Slough Estates in the operation of its business as a property development organisation. Because the organisation most commonly developed and retained the property for investment (leasing the properties to its clients), Slough had an unusually broad understanding not only of the construction process but of the process of development and the ongoing demands of property ownership and maintenance. Under these circumstances, improvements in value for the developer and tenants, and the associated continuous improvement programmes, became embedded in the way in which the organisation was managed. The term supply chain management was certainly not overused in the way that can occur elsewhere, and yet, Slough Estates provided a model of SCM in practice that others would have been well advised to emulate. Detractors will argue that Slough Estates frequently worked in a limited geographical area, within which they enjoyed privileged status with local Town Planners, that they frequently already owned the land that they developed and that they worked in a single, fairly specialised, market. The significance of the latter being that standardisation helps considerably in the effectiveness of supply chains and helps in improving value and reducing costs. Slough undoubtedly benefited from excellent knowledge transfer (see also **Chapter 4** above), typically from trade operative to senior management within the 'developer client' organisation and vice versa. Slough Estates is also among a small group of firms in UK construction that actively encourages the activities of postgraduate researchers within its organisation.

Potts charted the evolution of supply chains and the approach to managing those supply chains over a 15 year period at British Airports Authority (BAA). **Chapter 8** like the chapter that preceded it, focused on the experiences of one of the UK's largest construction clients; in the case of **Chapter 8**, a privatised airport operator, which manages the London airports along with a number of other airport management businesses in the UK and abroad. BAA has documented its initiatives through its in-house magazine 'In Context' and has been reported upon extensively in the construction and engineering press. **Chapter 8** reviews the management of the aftermath of the failure of the Heathrow Express tunnel, during construction, in 1994 and the way in which BAA worked collaboratively with the supply chain to problem solve and recover the project rather than descend into adversarial relationships and dispute. Later, the Genesis project case study demonstrated how, during the late 1990s, BAA devised a system of procurement and management, very much exploiting a SCM approach that was intended to be used on the Terminal 5 Heathrow project. Potts detailed how integrating the project team, mapping the supply chain, developing SCM, component-based design and a productivity improvement programme, transformed BAA's management of construction projects. The use of delivery teams associated with framework of suppliers was particularly important in terms of the innovative structure and processes employed. Potts emphasised the importance of mapping the supply chain and explains how this was used to great effect on the case study project, the clients particularly targeting the structural frame and mechanical and electrical (M&E) services for analysis. Finally, and to bring the BAA case study up to date, Potts looked at Terminal 5 Heathrow (T5), a project that was completed as this book went to print. The chapter covered the *M&E Buy Club*; the *Construction Logistics Consolidation Centres*; *3-D modelling with a Single Model Environment*; *Value Engineering*; *Offsite Prefabrication* and *Project Control Systems*. The chapter came to a close with a reference to the responsible and transparent transfer of risk under the T5 Agreement and one cannot but help feel that this is a key factor in successful SCM. We return to the subject of risk once more a little later in **Chapter 9**.

Much of what little has been written on SCM in construction tends to focus upon the client's perspective. **Chapter 9**, by King and Pitt, was written by a collaborative contractor/academic dyad and deals specifically with the contractors' position in relation to SCM. The chapter outlined a model that might serve as an alternative to that proposed in Holti *et al.* (2000). Essentially, the chapter took the position that the contractor is very well placed to manage the supply chain and that it is appropriate that the contractor does it on behalf of the client and in order to serve its own commercial interests. King and Pitt proposed SCM as a means of improving the project delivery process and questioned the conviction that only the client can successfully achieve this. The argument about contractors establishing standing supply chains and using these relationships to achieve competitive advantage echoed some of the points made by Skitmore and Smyth in **Chapter 5**. King and Pitt cited Mouritsen *et al.*'s (2003) work and they cautioned against the universal promotion of integration and collaboration without taking account of the supply chain environment and relative leverage positions of each of the supply chain actors. The concepts of project supply chain (PSC) and organisational supply chains (OSC) were discussed and the distinction was made between the PSC which relates to one client and the OSC which serves the contractor more generally (Male and Mitrovic, 2005). The existence of the OSCs, King and Pitt observed, provides the opportunity for main contractors to manage and influence a number of project-specific supply chains for a number of different clients, '. . . irrespective of the client's inclination and ability to utilise SCM'. King and Pitt used action research methods to focus upon five key aspects of issues in relation to supply chains:

- Relationships
- Culture
- Consolidation
- Consistency
- Cost.

The importance of maintaining and managing relationships was emphasised in conclusion, and anecdotal evidence was cited in support of this.

Smyth completed the contributions, his **Chapter 10** on franchising the supply chain raising a number of important issues about the practicalities of implementing SCM in construction. He put forward a model of franchising in construction – the idea that large subcontractors and the relatively smaller firm or individual can buy the rights to operate a well established brand. Smyth looked at the franchising idea from both the franchisor (brand owner) and franchisee (operator) perspectives. The franchisor overcomes the agency problem associated with running a relatively large firm through franchising. The franchisee gets all of the benefits of operating within a large firm, including training, marketing, quality control, codes of behaviour and administrative support. Smyth observed that construction has traditionally not been client-orientated. The procurement-driven client might logically be faced with a marketing response from contractors and yet contractors tend to pass on the procurement initiative to the lower tiers by merely demanding lower costs and higher added value (Smyth, 2005) and (Olayinka and Smyth, 2007). Smyth felt that contractors have shied away from incurring the marketing and procurement management costs of supply chain management (see also **Chapter 9** on the implications of this behaviour). In summary, under the franchising as one alternative, the franchisor (brand owner) gains:

- Resources, rapidly available over a wide geographical area;
- Risk spreading;
- Income from licensing of brand;
- Maintenance of service quality.

On the other hand the franchisee gains:

- Economies of scale not available to the small start-up firm;
- Reduction in start-up risks;
- Access to customer base and established reputation;
- A tried and tested product or service with established procedures and protocols.

The chapter is essentially a critique of the existing procurement models in construction. Franchising exploits relational contracting and secures continuous improvement alongside the effective control of product and service value for clients or end-users.

11.1.3   What does it all mean conceptually?

A number of contributors have questioned what is meant by supply chain management in construction. Perhaps if we reflect on the origins and motivations for the introduction of SCM in construction, we can see what it comprises. Construction needed a structure that provided collaborative relationships (see also Smyth and Pryke, 2008) and which maintained the flexibility demanded by the business environment. The potential loss of client level leverage implicit in partnering arrangements led the most forwarding thinking clients to think about how they might manage from the centre of the coalition, not only to maintain and improve value over the term of the relationship, but to promote improvement and innovation. SCM provides a means for managing the actors comprising the project coalition without the need to return to direct employment and management which has generally proved unsustainable, as evidenced in the British construction industry of the twentieth century.

Although the principle of managing firms with which a client (or other organisations acting in a SCM role) has no direct contractual relationship gains credibility and acceptance through lean production and logistics management, a number of contributors have questioned whether the term *chain* is appropriate. *Supply chain* is a reference to the sequence and contractual hierarchy through which construction is procured, but not to the way in which such supply chains are observed operating and managed. When we observe the functioning of the supply chain we observe networks of actors linked by a number of quite sophisticated relational linkages (Pryke, 2004, 2006).

This book challenges the traditionally held view that construction is fragmented, defined solely by organisational boundaries, and that fragmentation is a bad thing (see also Pryke, 2002). What is implicit in the account of SCM given by Rimmer (**Chapter 7**), Potts (**Chapter 8**) and Smyth (**Chapter 10**) is that increased fragmentation is frequently associated with the application of supply chain management. This fragmentation can be desirable, provided that risk is allocated fairly and that there is adequate compensation (see **Chapter 9**) and that quality systems, quality control training and procedures are maintained and reinforced. The upside of this fragmentation is shorter *path lengths* in communication terms. Previously highly specialised knowledge trapped within small specialist subcontractors and suppliers can now become available to clients and designers in a way that has not previously been possible. The barriers have been partly cultural (see Bresnen, **Chapter 4**) and partly a function of the structure and systems in place in construction.

SCM exists in the form of a chain at a high level of abstraction and it is the networks of relationships that provide us with the detail and analysis that we need to fully understand the operation of the supply chains. Yet the mission statement associated with the recognition of the importance of supply chains and their management is significant. By declaring an interest in SCM, we are moving on from the dyadic contract management and coordination management of the past. We are recognising that projects are achieved through people and that those people form individual and organisational relationships and contribute to business-to-business relationships as part of their daily working lives. The management of those relationships, using a supply chain approach (in another words escaping from the management of those firms in direct contractual relationships with the other firm) improves knowledge for academe and practice, thus contributing to the management of projects in construction.

11.1.4   What does it all mean for the industry?

The book has dealt with the SCM exploits of two of the largest and most successful construction clients in the UK. It has also proposed a model for franchising for those who would argue that SCM is only for the very largest clients – those with the resources, knowledge and inclination to pursue an SCM role from a very central position within the project coalition.

The exploits of Slough Estates emphasised the importance of abandoning a traditional hierarchical, contract-related structure for communications, in favour of non-hierarchical network of information and knowledge exchange freeing up, in particular, 'bottom-up' knowledge and information transfer. Slough Estates recognised before many that work or technology clusters had value *strategically*, as well as *operationally*.

Culture featured in several areas of the discussion, being seen as a limiting factor in the implementation of SCM. There was also a feeling that a change of culture within organisations to start thinking about value accumulation throughout the supply chain might provide a powerful route to innovation for clients and increased competitiveness for supply chains as commercial market actors.

Leverage (Cox, 2001), or relative power in supply chains, is an important factor in understanding the behaviour of supply chains. It affects the flow of information and knowledge throughout the network of actors that comprise the supply chain; leverage also has an important impact upon the way in which risk is transferred between supply chain members. Construction is an industry, within which risk is routinely transferred unfairly, in an opaque manner and without consideration for the ability and capacity of those receiving risk to effectively manage such risk. Risk is so frequently dealt with by those within the supply chain least able to defend themselves against such unfair risk transfer.

The concept of SCM places an emphasis upon strategy and value for clients and industry and a move away from a project task orientation. The potential is presented for long-term standing supply chains to develop over time and improve and innovate and in so doing provide better business solutions for clients, better project outcomes for stakeholders and higher levels of profitability for supply chain members.

Several contributors have, by implication, exploded the long term myth that fragmentation must of necessity be a bad thing for the construction industry, its clients and suppliers in all tiers. In some sense, a very small firm operating within a very competitive and innovative standing supply chain, linked to a well organised client, is better placed in terms of profitability and long term survival than a larger firm operating outside of such effective supply chains. It is suggested that it is the communication network path lengths that are more important to the industry, its firms and its clients, than the size of the firms. Short path lengths (the distance that information or knowledge must pass in terms of actors that will handle such material) affect the extent to which knowledge and information can travel, the quality of such material on arrival and the attitude that the eventual receiver might have to the material. If we move the emphasis of our management thinking away from *projects* towards *supply chains*, the issue of fragmentation becomes of lessening importance. The change of emphasis highlights the importance of the effectiveness, efficiency and ultimately dominance of supply chains within the market place, whether it is domestic or global.

The concept of supply chains and their management as supply chains, help us to assemble groups of suppliers and contractors and to manage them in a way that places emphasis on value and cost and to understand that the group of project actors might collaborate to share information and knowledge and to share and manage risk in a manner that is equitable and transparent.

11.2   Final Thoughts and the Future of SCM in Construction

The value of SCM lies in the conceptualisation of construction systems through the structure of supply chains. Huge benefits await industry and researcher alike in the mapping and analysis of supply chains in pursuit of both added value and cost reduction. Although there is some evidence of activity in this area, it is limited and much greater exploitation is potentially possible. The key to dealing with such mapping and analysis is the solution of the technological and social barriers preventing the integration of construction supply chains through the application and rigorous exploitation of IT in this area. There are some groundbreaking examples of innovation in this area, but the construction industry as a whole needs motivating and educating.

Part B

Part B

At present trust and relationship management are regarded as a substitute for the contractual governance of construction systems. These human and contractual aspects need not be mutually exclusive. Some work has been done on devising appropriate contract forms (PPC 2000, amongst others) but there are still a number of contractual issues associated with other aspects of relational contracting (see Pryke, 2006a), which need further research and resolution by the industry.

Supply chains and clusters within such supply chain networks provide the structural format for some interesting innovations in the construction industry. These structures have been perceived in the past as structures enabling innovation in essentially *operational activities*. This needs extending to the strategic management of construction firms, project coalitions and the industry as a whole. Supply chains managed through effective networks of relationships and supported by effective financial incentives might provide a vehicle for radical shifts in value produced and profitability achieved.

11.3   In Conclusion

Construction is essentially, for the most part, complex and profitability tends to be low; long chains of command (or path lengths, see above) provide the industry, the project actors and the industry's clients with convoluted communication links and unfair risk allocation practices. The risk minimisation/avoidance strategies adopted, quite commonly by actors at all levels, lead to an environment where innovation is difficult to nurture and is generally quite rarely seen. SCM provides the industry with an opportunity to understand the process of assembling the materials and components necessary to deliver *customer delight* (Latham, 1994); to map and understand how both cost and value accrue over the length of the supply chain and to begin to understand how recognising the supply chain, giving it an identity and harnessing the strengths of such supply chains, provide the potential for reform and improvement within the industry. Indeed some might argue that a failure to identify and harness the strengths within the supply chains in construction will eventually lead to superior competitive strength from an industry that is starting to feel global as a relative latecomer.

11.4   Next Steps

Increasingly, competition in construction will be based upon the perceived potential for innovation and value creation, rather than a simple lowest capital cost bid. These drivers will give an intense focus to the supply chain as a source of such innovation and value creation and highlight the limitations of the project as a structure within which innovation and value can be driven up. It is for the academics to lead on the identification of methods for the analysis of supply chain activity and it is for the practitioners to apply these methods and learn more about the nature and features of the supply chains that are so important to their future competitiveness.

References

Cox, A., Ireland, P., Lonsdale, C., Sanderson, J. and Watson, G. (2001) *Supply Chains, Markets and Power: Mapping Buyer and Supplier Power Regimes.* London: Routledge.

[Egan report] DETR (1998), *Rethinking Construction: The Report of the Construction Task Force to the Deputy Prime Minister, John Prescott, on the Scope for Improving the Quality and Efficiencies of UK Construction*, DETR at www.construction.detr.gov.uk/vis/rethink

Elmuthi, D. and Kathawala, Y. (2001) An overview of strategic alliances. *Management Decision*, 39(3), 205–217.

Gray, C. (1996) *Value for Money*, Reading Construction Forum and the Reading Production engineering Group, Berkshire.

Green, S.D., Fernie, S. and Weller, S. (2005) Making sense of supply chain management: a comparative study of aerospace and construction. *Construction Management and Economics*, 23, 579–593.

Hofstede, G.H. (1994) The business of international business is culture. *International Business Review*, 3(1), 1–14.

Holti, R., Nicolini, D. and Smalley, M. (2000) *The handbook of supply chain management*, CIRIA, London.

Latham, Sir M. (1994) *Constructing the Team: Joint Review of Procurement and Contractual Arrangements in the United Kingdom Construction Industry*, HMSO, London.

Male, S.P., and Mitrovic, D. (2005) The Project Value Chain: Models for Procuring Supply Chains in Construction. *QUT Research Week Conference, 4th–8th July, Brisbane, Australia: RICS*.

Mouritsen, J., Skjott-Larsen, T. and Kotzab, H. (2003) Exploring the contours of supply chain management. *Integrated Manufacturing Systems*, 14(8), 686–695.

Olayinka, R. and Smyth, H.J. (2007) Analysis of types of Continuous Improvement: demonstration projects of the Egan and post-Egan agenda. *Proceedings of RICS COBRA 2007*, 6–7 September, Georgia Institute of Technology, Atlanta.

PPc 2000 (2000) *Project Partnering Contracts*, Association of Consultant Architects, Bromley, Kent.

Pryke, S.D. (2002) Construction coalitions and the evolving supply chain management paradox: progress through fragmentation, RICS COBRA conference, Nottingham Trent University.

Pryke, S.D. (2004) Analysing construction project coalitions: exploring the application of social network analysis. *Construction Management and Economics*, 22(8), Oct, 787–797.

Pryke, S.D. (2006) Towards a social network theory of project governance. *Construction Management and Economics*, 23(9), Nov, 927–939.

Pryke, S.D. (2006a). Legal issues associated with emergent actor roles in innovative U.K. Procurement: Prime Contracting case study. *Journal of Professional Issues in Engineering Education and Practice*, American Society of Civil Engineers, Vol. 132, No. 1.

Pryke, S.D. and Smyth, H.J. (2006) *The Management of Complex Projects: A Relationship Approach*, Blackwell Publishing, Oxford.

Part B

Schein, E.H. (1990) Organisational Culture. *American Psychologist*, 45, 109–119.

Schneider, W. E. (2000) Why good management ideas fail: the neglected power of organizational culture. *Strategy and Leadership*, 28(1), 24–29.

Smyth, H.J. (2005) Procurement push and marketing pull in supply chain management: the conceptual contribution of relationship marketing as a driver in project financial performance. *Journal of Financial Management of Property and Construction*, 10(1), 33–44.

Smyth, H.J. and Pryke, S.D. (eds) (2008) *Collaborative Relationships in Construction*, Wiley-Blackwell, Oxford.

Index